Table of Contents

Introduction

Thinking in systems may sound like a distant yet appealing concept to the layman. It has a smart, technical, objective, and superior cognitive air to it, right? However, when I ask my students "What do you think 'thinking in systems' means?" they are usually confused and can't truly articulate a one-sentence answer.

I can't blame them. Because there is no articulate, definitive answer to this question yet. Experts work hard to give a comprehensive definition to systems thinking but even those summaries that get close to perfect are missing some essential parts. The term has been defined and redefined across the decades but no one has been able to press this intangible concept into a statement that will allow it to be measured.

Thus I can't give you a one-sentence answer either to what exactly systems thinking is, but I will provide you with a deep understanding through a multilayer analysis.

The dictionary defines a system as a group or an arrangement of things that work towards a common goal. Wikipedia says that a system is "a group of interdependent items that interact regularly to perform a task." We can state that the grounding principle of a system therefore is something more than a collection of its parts.[i]

A system is composed by parts that we call *elements*, which are *interconnected* to serve *a purpose or function*.

Let's take a farm as an example. A farm is a system which has the field, workers, seeds, machinery, and irrigation as elements. The different relations between these elements show how they are interconnected. They form, or are organized, into these interconnections for their overall function: to produce wheat, for instance.

A motorbike is also a system. This is a mechanical system with a number of elements like the engine, wheels, brake, lamps, and stickers, which, interacting together, serve the function to work as a unit of transportation.

Let's take a look at a natural system: a plant cell. This is a biological system. A plant cell is a mixture of organelles that are interconnected to perform metabolic processes. This enables the cell to function as an entire system.

If we take a more distant look on our three systems—the farm, the motorbike, and the plant cell—we can see that these independent systems are part of larger systems. On a small or large scale, these sub-systems affect the larger system above them, and conversely, the large system has definite influence over the smaller subsystem.

For example, the farm belongs to a system of regional economy. Although the whey production of one farm may not have significant impact on the overall economy of a country, it matters. At the same time, economic fluctuations, price changes, and the shifting of supply and demand can have a large impact on the farm.

The motorbike belongs to a larger system, too. For example, the system of local traffic. One motorbike doesn't affect the larger system too much if it functions well, delivering its rider safely from A to B. However, if the motorcycle rider has an accident which causes a blockage on the road and a heavy traffic jam, it has a temporary, yet significant impact on the system. The traffic system has an even greater impact on the motorcycle driver in forms of all-time regulations, a speed limit, an age requirement, maintenance conditions, etc.

The plant cell, while it is an individual particle, works together with other individual cells to sustain the function of the larger system, let's say a flower. The combined effort of the particles helps the plant survive, gain nutrients, photosynthesize, and grow.

All of the big systems mentioned above, the regional economy, the local traffic, and the flower belong to even bigger systems, and so on. No system is independent; none of them live in isolation. They are interdependent.

Following the thought thread of system interdependence, we can say that systems thinking is "a system of thinking about systems."[ii] Using the three parts of systems, elements, interconnections, and function or purpose, systems thinking allows us to:

- "Understand how the behavior of a system arises from the interaction of its agents over time (i.e., dynamic complexity);
- Discover and represent feedback processes (both positive and negative) hypothesized to underlie observed patterns of system behavior;
- Identify stock and flow relationships;
- Recognize delays and understand their impact;
- Identify nonlinearities;
- Recognize and challenge the boundaries of mental (and formal) models;
- Recognize interconnections;
- Understand dynamic behavior;
- Differentiate types of flows and variables;

14

- Use conceptual models;
- Create simulation models;
- Test policies;
- Incorporate multiple perspectives;
- Work within a space where the boundary or scope of problem or system may be 'fuzzy';
- Understand diverse operational contexts of the system;
- Identify inter- and intrarelationships and dependencies;
- Understand complex system behavior; and most important of all,
- Reliably predict the impact of change to the system."[iii] (Ross D. Arnold, 2015)

This book focuses on identifying nonlinearities and analyzing the dynamic behavior of complex systems. I will introduce the concept of chaos, where I will show how small differences in the way things are now can bring great consequences in the way things will be in the future. We will explore the phenomena of chaotic behavior, taking a closer look at the Butterfly Effect, bifurcations, phase transitions, and fractals. For transparency and easier understanding, we will focus on systems related to science and mathematics.

In these fields we distinguish two types of systems: linear and nonlinear. Let's take a closer look at what they are.

Chapter 1: Linear Systems

What are linear systems?

Linear systems obey certain rules; they are defined by their adherence to what is called the superposition principle.

A quick definition of the superposition principle sounds as follows: "The net response caused by two or more stimuli is the sum of the responses that would have been caused by each stimulus individually. If input A produces response X and input B produces response Y then input (A + B) produces response (X + Y)."[iv]

In more simple terms, if we have two or more inputs at a given point in time, the final output will be the result of adding all the outputs.

Establishing all of the scenarios in an input-output system using infinite measurement is practically impossible. However, when the system in question qualifies as a *linear system*, one can use the reactions established through a base set of inputs to forecast the responses to other possible inputs. Doing this saves a lot of work and makes it possible to predict and identify the system.[v]

How can we identify a linear system? Well, as I said before, we need to see if the system adheres to the components of the superposition principle, namely additivity and homogeneity. Let's see what they are.

Additivity principle

The additivity principle states that we can add the output of two systems together and the outcome of the systems combined will be the addition of each individual system's output in isolation.

For example, if I had two oxen that could each pull 300 lbs. of cargo on a cart in isolation, when I combine these two oxen to pull a larger cart they will each be able to pull twice as much weight: 600 lbs. In mathematical terms the additivity principle looks like this: 300+300=600.

The additivity principle therefore can be explained with the terms we used to explain the superposition principle: "The net response caused by two or more stimuli is the sum of the responses which would have been caused by each stimulus individually."

Homogeneity principle

The homogeneity principle states that the output of a linear system is always directly proportional to the input.[vi] In other words, if we put twice as much into the system, we'll get twice as much out. In numbers, the homogeneity principle looks like this: 1x=2, 2x=4, 3x=6, 4x=8, etc.

In real life terms, if you paid $40 for a wine from which you expected a certain quality, the principle states that if you paid twice as much ($80) you would get a wine that was twice as good.

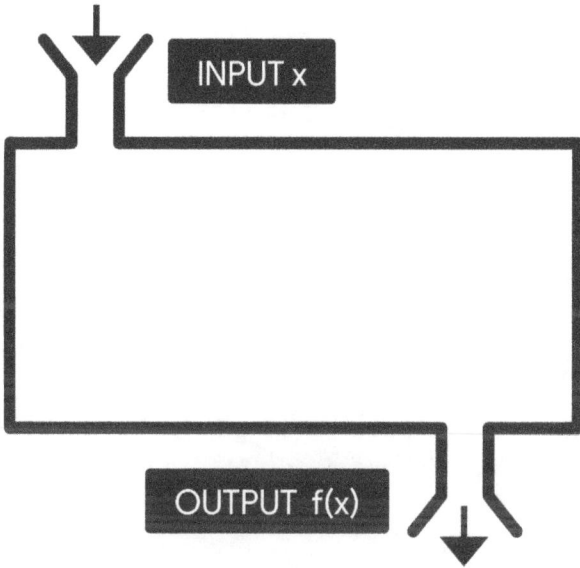

Picture 1: Linear Systems

If we put this on a graph, we would see why a linear system is called linear: the result is always a straight line. Let's refresh our math knowledge with a simple example.

Let's solve a linear system by graphing. The point in which the two equations intersect will

be the solution to the system. For example, let's
take these two equations:

$$y=2x+4$$
$$y=3x+2$$

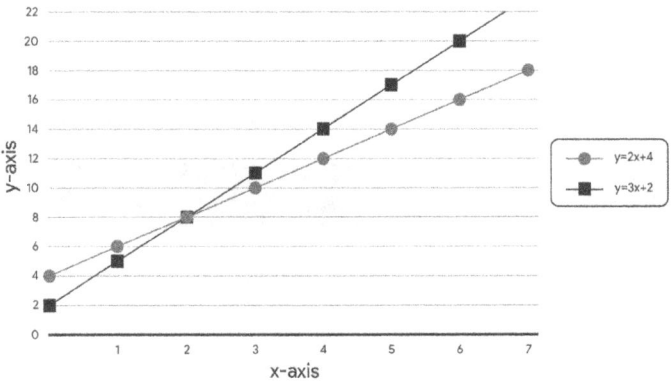

Picture 2: X-Y Axis

These two lines seem to intersect at 2 on the x-
axis and 8 on the y-axis. Let's see if this is true
algebraically. It never hurts to double check the

graphic solution. Also, when the equation is more complicated, it's necessary to calculate the results first and then illustrate them.

$y=2x+4$ ➔ $y=2 \cdot 2+4$ ➔ $y=8$

Let's check it on the other equation, too.

$y=3x+2$ ➔ $y=3 \cdot 2+2$ ➔ $y=8$

A linear system that has only one solution is called a consistent independent system, where consistent stands for the lines intersecting and independent stands for the lines being distinct.

There can be linear systems of parallel lines that have the same slope but a different y-intersect. These linear systems don't have a solution as the lines never intersect. They are called inconsistent systems.

Linear systems where the lines have the same slope and y-intersect are called consistent dependent systems. They have infinitely many solutions as the lines coincide.[vii]

The additivity and homogeneity principles are deeply intuitive to us. On a basic, mathematical level they could appear very simple, but they assume a lot of given facts about how the world works.

First, let's analyze the guiding assumptions in support of the theory of linear systems.

The homogeneity principles (additivity and homogeneity) state that the isolated properties of a system are what matter. They disregard the way in which these properties are joined or the relation between them. How does this look in a real life example?

The case of the additivity principle

Let's say you want to fertilize your garden and there are two fertilizers that are meant to solve your garden problem. You buy them and use both at the same time. In this case the output or result of this system will depend on whether the two fertilizers have an effect on each other. If they have no effect upon each other, it will be the properties of each fertilizer in isolation that will determine the output of the system (the quality of your garden's soil). There would be a lack of interaction between the two elements (fertilizers) in the system, thus the linear model will be able to fully capture this event.

However, if the fertilizers affect each other, it will be the interaction between them that will determine the system. The linear model fails in this case as it relies upon the additivity principle that connotes an additive relationship.

But in this situation simple additivity is not the case.

The basic reasoning of the additivity principle is that we can add elements together without taking into consideration how these elements interact together. Our example of fertilizers proves how frail linear systems theory is in the grand scheme of systems thinking. But before jumping to conclusions, let's test the homogeneity principle, as well.

The case of the homogeneity principle

The principle of homogeneity assumes that the scale doesn't matter. Is that so, though, in case of real world systems? Let's say you have a store where you import one million buttons a year to sell. Let's assume that you are successful and you can sell all of your buttons by the end of the year. Now let's scale up the import to two million buttons a year. If everything would be scaled in a linear fashion, the linear model could capture this. But, of course, in the real world chances are very low that all the other variables would scale according to the laws of linearity. Would your costs scale in a linear fashion? What about market saturation? If you can't sell the two million buttons your revenue will not grow linearly either.

The key takeaway of these examples is that linear system models can't capture feedback. They fail to consider the effect the actions of a system has on its environment. Linear system models also lack the potential to show how the environment will in turn affect the system; not only in space, but also over time. In other words, how will past decisions and actions feed back to affect the present conditions of a system.

Why do we talk about and use linear systems then?

Linear system analysis happens in a static time vacuum but there are a few reasons why we use linear systems.

- Linear systems are intuitive. The static properties of tangible elements and events that linear systems capture are easier for us to understand and quantify as opposed to the elusive world of interactions between elements and their behavior over time.

- Linear systems can catch the behavior of a few systems. For instance the interactions between the simple dynamics of cause and effect happening in social and economic behavior.

- Linear models are simple. They exclude qualitative questions and analysis about the interactions between the elements in a system. They are well-suited to the strict quantitative methods of mathematics and the reductionist approach of science. Thus we can approach complex problems by breaking them down and then tackle the more simplified problems in isolation.

Using linear systems to help with real life situations

The good news about linear systems, in a nutshell, is that if you pick the right set of inputs you will know how the system behaves under any condition. This is because superposition allows you to forecast how your system will work, even under circumstances that you did not measure. This is the core of linear systems analysis.

The bad news is that purely linear systems are hard to find in the real world. There are, however, systems that have a range of inputs which react linearly, and as long as you work within that range your system will work.[viii]

Linear systems are not only for scientists, mathematicians and engineers. We may not realize it, but many of us are already using linear systems when we compare the costs and benefits of a variety of purchases.

For example, we need to establish if it is cheaper to buy an item online where we have to pay for shipping, or at the store where our only additional cost is the commute to the store. It may seem like a very simple problem, but without you even realizing it you used the theory of linear systems to ascertain which would be the most effective method to come by your new purchase.

Let's look at a tangible example published in Algebra.[ix]

Dianne is trying to choose between two internet providers. The first plan with T-Mobile would cost $20 per month for a set amount of data with an additional 25 cents as cost per every additional MB (megabyte) she uses. Her second option, Verizon, is $40 per month, but their cost per additional MB is only 8 cents. Which should she choose?

By now we should feel intuitively that the choice will depend upon how many extra MB she expects to consume monthly.

Let's write down the two equations for the cost in dollars in terms of the MBs used. The number of MBs is the independent variable; that will be the x. The cost is dependent on the

MB. Thus the cost per month is the dependent variable, the y. Our equations will look like this:

- For T-Mobile: $y = 0.25x + 20$
- For Verizon: $y = 0.08x + 40$

Let's write the equations in slope-intercept form ($y = mx + b$), and illustrate them as a graph:

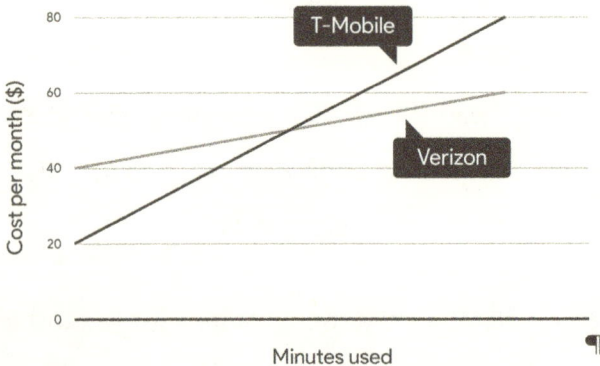

Picture 3: T-Mobile –Verizon Graph

The plan for T-Mobile has an intercept of 20 and a slope of 0.25. The Verizon plan has an intercept of 40 and a slope of 0.08. To help Dianne make an informed choice, we need to find where these two lines cross. Let's solve the two equations as a system.

Equation T-Mobile gives us an expression for $y(0.25x + 20)$, so we can substitute this expression directly into equation Verizon:

$0.25x+20=0.08x+40$ → subtract 20 from both sides

$0.25x=0.08x+20$ → subtract 0.08x from both sides

$0.17x=20$ → divide both sides by 0.17

$x=117.65$ MB

The numbers are not pretty but we can conclude that if Dianne uses an extra 117.65 MB a month, the two plans would cost the same. Now let's go to our graph to assess which plan is better if she uses more than 117.65 MB and which is better if she uses less. We can see that the T-Mobile plan costs more is she uses more than 117.65 MB and the Verizon plan costs more if she uses less than 117.65 MB.

In conclusion, if Dianne uses 117.65 MB or less each month she should go for T-Mobile. If she estimates she'll use 117.66 MB or more she should take Verizon's plan.[x]

As you can see, simple linear equations can be useful for minor everyday decisions.

In conclusion

Science is everywhere, and even if you are not aware of it, you are using linear systems to solve everyday problems. It's not necessary to know the mathematical formula, but it is nice to know that your brain works that way.

Chapter 2: Nonlinear Systems

What are nonlinear systems?

By definition a nonlinear system "is a system in which the change of the output is not proportional to the change of the input." [xi] Nonlinearity is an important part of mathematics and science with a special interest to engineers, biologists, physicists and other fields of science that are nonlinear in nature. [xii] Dynamic systems are nonlinear most of the time and through them we can analyze changes in variables over time, which may seem chaotic, unpredictable or counterintuitive. [xiii]

Nonlinear systems set out almost all real world phenomena yet they have been assigned the role as alternatives to scientific or mathematical analysis as opposed to linear thinking. Why?

Because linearity was and is easier to model. From a scientific viewpoint, we could only understand the world through our models. Setting up a model for a two-variable system is inherently easier than capturing all the influencing variables related to an event. It is almost impossible. Thus even though our world is complex and operates by the laws of nonlinearity, scientists simplified their tools of understanding so that they at least have one tool to use. When I say simplified, I mean the description of phenomena that are the result of cause and effect interactions. Simple interactions allow us to connect the cause and effect with a simple line; in other words, in a linear fashion.

Scientific focus has been grated largely to simple, linear interactions in the past centuries;

organized geometric forms described in neat equations were the alpha and omega of mathematics and science for the simple fact that they are easy to encode in scientific language and understanding.

As the fields of science developed, the models through which they examined the world also became more sophisticated. They offered a wider, more complex representation of reality. The current approach to nonlinearity is the byproduct of the past few decades only. As our world expanded, sped up and became undeniably interconnected, scientists began to see and define the systems operating our world for what they are: nonlinear dynamic systems.

Let's take a closer look at how nonlinear systems are different than linear ones.

Nonlinear systems defy the superposition principles

Let's recall the main features of linear systems discussed in Chapter 1. These features are called the superposition principles, additivity and homogeneity. To distinguish a linear and nonlinear system thus is easy: systems that can be defined by additivity and homogeneity are linear and those systems that defy these principles are nonlinear.

Additivity

Remember, this is the principle that says when we add two or more elements the resulting combined system will be a simple addition of each element's properties in isolation.

We can't apply the additivity principle to nonlinear systems because the elements we put together and their interaction matter. Their relationship result in an outcome where the elements' combination is more or less than each element's properties in isolation. The rules of additivity thus do not apply.

For example, let's put two living beings together. Depending on what kind of being we choose, their interaction will result in qualitatively different kinds. If we take a flower and a bee, an impala and a leopard, or a gazelle and a gnu, their interactions will be significantly different. Bees and flowers have a synergistic interaction, the impala and the leopard have a predator-prey dynamic and the gazelle and the gnu will be competing for the same patch of nutritious grass.

This part of nonlinear interactions is easy to follow and understand. It's part of experiencing the world as it is and having common sense. The difficulty arises when we try to translate these phenomena into the language of mathematics and science. In these fields we are used to study the qualities of the elements in isolation—in a laboratory, for example—or in an otherwise separated environment. Looking through the narrow lenses of microscope (sometimes literally, sometimes figuratively speaking), it is hard to know when, why and how these elements would naturally interact with each other and if there is a special, surprising or new angle of emergence analysis in such cases.

It's also hard to effectively formulate a comprehensive language in math and science because we may not know how the elements will behave. For instance, imagine yourself trying to recommend a friend as a possible employee candidate to another friend. You know both of your friends, what they are like as individuals. But it is much harder to predict how they will get along in a work environment even if in isolation they seem like a good match.

Homogeneity principle

The homogeneity principle states that the outputs of a system are always directly proportional to the input. In other words, twice as much in results in twice as much out. This principle, however, can't predict the effect of the previous states of the system on the outputs. Often previous states have a significant effect on present or future conditions. In other words, linearity can't capture feedback; inputs and outputs come and go without any connection between them.

Systems operating in space and time will inevitably be affected by feedback loops from their environment as every action the system takes will drive some sort of consequence. This consequence will feed back to the future state of the system that in return will elicit a reaction from the system. Therefore, as soon as we position our system into the real world, things unavoidably turn nonlinear. The more feedback loops we capture and include in our system mapping model, the more realistic of a picture we get about reality—and the more nonlinear the world becomes.

Let's look at a practical example. In the heyday of the industrial revolution the paradigm was highly linear. The more you produce, the richer you become, the more developed the world will be. More trains equaled quicker and more broadened transportation and travel possibility. More factories equaled more goods to sell and so on. Nobody cared about the possible consequences of such sudden and significant changes, how the increased smog and garbage would affect the environment. Or how will environmental changes affect future industrial input and limits in return? The more environmentally conscious we become the clearer we see that growth and industrial expansion has a limit within our ecosystem. Endless scaling is not possible. The input will inevitably stop being directly proportional to output thus the laws of homogeneity won't apply anymore. We are dealing with a nonlinear system again. (The phenomenon of industrial expansion can be well illustrated in the framework of the Limits to Growth systems archetype.)

Superposition principles stop applying as soon as we take into consideration the interaction between elements within a system and outside forces (feedback loops) that affect the system.

Synergies and interference

If the principles of additivity and homogeneity are defining linear systems, synergies and interference can be the definer of nonlinear systems. Let's take a look at them.

Synergies

The word synergy originally means working together and comes from the Greek language. Synergy is a positive interaction between two elements derived from synchronization between their conditions. The synergistic forces create a net result that is greater than the product of the individual elements' actions in isolation.[xiv]

The division of labor, for example, within ant colonies could be a good example of synergistic interactions. The ants have different functions within the colony. Thanks to their division of labor and collaboration, their synergy creates a colony that exceeds the potential of the sum of each ant in isolation.

Two things have to be therefore aligned to create a synergy: differentiation (in our case the division of labor) and synchronization (in our case the cooperation between different ant functions). Imagine each ant having the same function; in this case their joint effort would be a simple addition. Remember the oxen each being able to carry 300 lbs.—it would be the same thing. Performing different functions without coordinated cooperation would again result in an additive linear system where we add apples with apples and oranges with oranges; forager ants with forager ants and builder ants with builder ants. Without coordination these functions wouldn't result in something greater.

To have synergic nonlinear systems, we need element differentiation and coordination. Differentiation should happen in a purposeful, intentional fashion, not as a result of random events. The differentiated elements then purposefully need to coordinate their actions.

Let's see another example. Let's say person A and B want to buy a jar of peanut butter which costs $10. However, each of them has only $5, which in isolation is not enough to buy the peanut butter. However, if they combine their money, they will have the $10 needed to buy the jar. They both differentiate their actions towards each other and then cooperate in order to achieve a common goal which creates synergy. With $5 each, they don't get any peanut butter. But once they add their fortunes together, the outcome is different and they both can enjoy their delicious peanut butter treat.

Interference

Synergies are constructive by nature; through differentiation and coordination they bring something more into the world. We can see this phenomenon playing out in the economy, social relations, the ecosystem, etc. But, of course, where there are constructive forces, there must also be destructive ones. These destructive forces are interferences.

Interferences as a consequence create a combined system which is less than the sum of its components in isolation. Think about two colliding waves which extinguish the power of each other, or the light reflecting on an oil patch floating on water separating the color spectrum. [xv]

As opposed to synergistic relations, interference often lacks the pivotal differentiation between the elements within the same environment. This often plays off in each element trying to access or control the same state at the same time. The result will be a destructive competition between them where everyone will inevitably gain less.

A simple illustration of this is rush hour traffic where many cars are competing for the same route at the same time. Each new car will slow down traffic even more, causing precious time loss for every participant due to a lack of—in this case route—differentiation.

Synergies are adaptive and area able to synchronize their behavior with other elements of the environment thanks to the positive feedback mechanisms from which they arise. Interference, on the other hand, lacks feedback between the components, thus it's not adaptive and also won't be able to synchronize behavior with other elements in the environments. Interference systems are asynchronous.

A living organism is a complex system that emerges out of differentiation and synchronization. This happens on multiple levels at the same time; on a cellular level, on the level of organs and so on. They are operating through a consistent network of feedback loops. Cancer, for example, is a state where cells defy differentiation and synchronization and start acting freely without obeying the body's control mechanisms. They grow into a harmful tumor which then destroys the relation of all the other parts of the living organism's system.

Chapter 3: Feedback Loops and Equilibrium

What are the feedback loops?

The model of a linear system would include an input, some process and an output. The input and the output of such a system are independent of each other; the value of the current input is not affected by the previous output, nor will the next one be affected by the current input.

This is a model which holds true in certain cases. For example when we flip a coin; just because we flipped tails ten times in a row, it doesn't mean that chances are now higher to flip heads next. Chances are always fifty-fifty and previous flips have no effect on the next one. We call this phenomenon the Markov property where the conditional probability distribution of future states of the process (conditional on both past and present states) depends only on the present state, not on the chain of events that preceded it.[xvi]

However, most things in the real world don't work in accordance with the Markov process. In real life the present input variables depend a significant amount on previous outputs. Similarly, the output of today will have an impact on the input of tomorrow. For example, the state of the economy today will affect the state of the economy tomorrow. The weather today affects the weather of tomorrow and so on.

The condition where the output of a system has an effect on the input as a part of a chain reaction of cause and effect is called a feedback loop. The "what goes around comes around" saying applies in the case of feedback loops; an effect is returning to its cause and creates more or less of the same effect. For example, when you have an argument with someone, that will affect how you feel about and talk to the same person the next time you meet (if the conflict is not resolved).

There are two types of qualitatively different feedback loops.

Negative or balancing feedback loop

A balancing feedback loop constitutes a connection of constraint imbalance between two or more variables. In other words, when one of the variables in the system turns in a negative direction another variable in the system turns in a positive direction. The maintenance of the overall combined value of the system is the goal in such a dynamic.

Think about a thermostat that is set to keep the room at a certain temperature, let's say 70 F. Whenever the temperature of the room drops below that setting the furnace turns on to counterbalance the heat difference.

Or think about the general economic law of supply and demand. If there is an increased demand for a product, prices may rise until the product becomes so expensive that demand decreases for it.

It may seem that there is a simple additive relationship in these cases; one variable goes up and as a consequence the other one goes down, proportionally seeking equilibrium as a result.

Equilibrium has a crucial role in the linear systems theory as it produces a zero-sum game. The combined value of gains and losses is zero. The system can be defined by this feature; equilibrium being the system's normal state.

Having this information, we can build rectifying models around it if the system faces imbalance.

Unfortunately, in real life this rule can only be applied to systems in isolation that are affected by balancing feedback mechanisms. In the nonlinear world the assumption of equilibrium as a normal state crashes.

Positive or reinforcing feedback loops

In a reinforcing feedback mechanism the value-increase of one element is correlated with the increase in values of another element. In simpler terms, both elements either grow or decrease simultaneously. The result of a process is always self-reinforcing which leads to exponential growth or decay. The gains and losses are not additive and they don't sum up to zero. There is no equilibrium in reinforcing feedback mechanisms; the system is far from equilibrium as a result.

Take compound interest as an example. Last year's interests affect this year's total amount of money. The more money you have in your account, the more interest you'll earn next year.

Reinforcing feedback mechanisms are unsustainable, however. They need energy as input from their environment. Remember the example of the unsustainability of the industrial revolution? Think about the development of petroleum processing technologies. The easier and cheaper we access and process petroleum the more we feed back in energy and thus in industrial development. There will be a point, however, where we will hit some limit; for example, we exploit all of Earth's petroleum reservoirs.

Understanding the dynamic of balancing and reinforcing feedback loops are crucial to comprehending nonlinear systems.

When we take the system out in the real world, the system's output will affect its environment. In return the environment with some time delay will affect the system's future inputs. If the future input will have lead in the opposite direction as the previous output we talk about a balancing feedback that stabilizes the system, seeking equilibrium. If the future input will accelerate the dynamic of the system in the same direction as the previous output, we're talking about a reinforcing feedback loop which creates exponential growth or decay moving away from the state of equilibrium.

Chapter 4: Exponentials and Power Laws

What are exponentials?

The ratio of input and output does not fluctuate in linear systems. However, when we add feedback loops to the picture, the past state of the system affects the current state in the form of feedback. This process enables the present ratio between the input and output to be bigger or smaller than the previous ratio.

This qualitatively different event is called an exponential. In mathematical terms exponentiation is written as b^n. These letters stand for two numbers, the base b and the exponent n. When n is a positive integer, exponentiation corresponds to repeated

59

multiplication of the base: that is, b^n is the product of multiplying n bases:[xvii]

$$b^n = \underbrace{b \times \cdots \times b}_{n \text{ times}}.$$

Picture 4: Exponentials

The exponential symbol shows the process of taking a variable and times it by another variable, not only once but rather in iteration. Iteration is the repetition of a process in order to create a chain of outcomes. The progression will approach some end value or goal. Each repetition of the process is a single iteration. The output of each iteration is then the starting point of the next iteration. [xviii] In systems thinking terms we take an output and feed it back into the system calculating the next output.

Let me give you a practical example. I'm a chemist and I wish to build a test tube of bacteria where the number of bacteria will double every second. I begin this project at noon with only two bacteria. The plan is to have the tube full by early evening. As seconds pass, the population will grow exponentially, as the number of bacteria at the time will feed back and affect the population of the next second.[xix]

$b^n = b \cdot b$

$2^2 = 2 \cdot 2 = 4 \rightarrow 4^2 = 4 \cdot 4 = 16 \rightarrow 16^2 = \ldots$

As you can see in the calculus above, the number grows exponentially, however, it will take a few hours of exponential growth and the test tube will still be only 3% full. However, in the next five seconds, as we get closer to evening in time, the test tube will get fully filled.

This disproportional, extreme change is quite counterintuitive to linear thinkers. It doesn't follow the natural, step-by-step proportional linear progression. When we face exponential growth, we can see that the rate of growth is also growing, not only the variables. It's not hard to see how crucial and dramatic changes can both exponential growth and decline create.

What is power law?

We can also call exponentials as powers. Power law depicts a functional connection between two quantities where one quantity changes as the power of the other. Imagine scaling an object like a cube. A cube with "X" side lengths will have a volume of eight cubed. The number by which we are multiplying the volume thus grows every time. If the volume was twice the size's length, we would be talking about a basic linear scaling.[xx]

Metcalfe's Law

Metcalfe's Law comes from telecommunications and later IT networking. "Metcalfe's Law is attributed to Robert Metcalfe, a former researcher at the famous Xerox Palo Alto Research Center who co-invented ethernet and it states that the value of a telecommunications network is proportional to the square of the number of connected users of the system (n^2)."[xxi] In other words, it is based on the observation that the number of possible cross connections in a network grows as a square of the number of phones in the network. Each time we add a new device into the network we can add as many more links as our devices already in the network.[xxii]

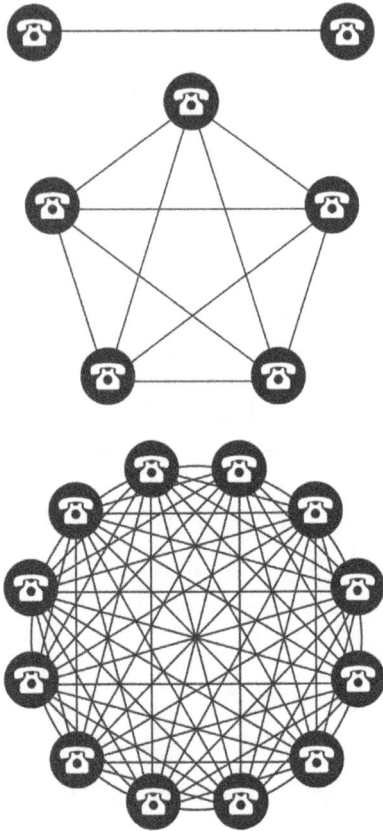

Picture 5: Metcalfe's Law

By obtaining a phone and a number we can access the paid phone network and connect to another phone and number anywhere in the world. The value of the network for users increases exponentially over time with every phone added to the network. As we can see, the number of phones as communication devices grows in a linear way in the network. However, the number of connections grows in an exponential way, making the network more powerful and valuable for every existing user. Metcalf's Law improves the power of the network by increasing proportionally the square of the number of devices in the network.

This law applies to all kind of networks, therefore we could call it the network effect, which is the key of reinforcing feedback. Why? Because each time someone connects with another element in a network, it becomes more likely that someone else will, too. The example of the expanding telephone system presents fairly well the dynamics behind reinforcing feedback mechanisms which can force the system towards growth or decline very quickly. Yes, we shouldn't forget that a nonlinear system can create exponential change in two directions, growth and decline. While today it seems fairly impossible that the telecommunication network would start to decline exponentially, in theory that can be a possibility. Also, by having a widespread networking system such as the internet, bad news can go viral in seconds and can cause greater distress than indicated.

Long tail distributions

Exponential power law in nonlinear systems is not normal. Let me explain what I mean. They have neither statistically normal averages nor archetypical states within the system. To understand more in-depth what I'm talking about, let's take a look at what is average and normal.

Normal distribution

Normal distribution implies that the system has a mean state once we collect enough sample states within this system. We can thus calculate an average that we can use as the classic state of the system.

Let's take shoe sizes as an example. If I organized 1000 random men based on the size of their feet, this sorting would follow the law of normal distribution. Meaning, there would be a lot of people who have sizes 9-11. And fewer others who have either bigger or smaller sizes. The normal distribution has a well-defined middle and from there the other numbers (shoe sizes) drop exponentially. In other words, chances are extremely low to get someone with a shoe size on either extreme.

We could position this normal distribution on a Bell Curve:

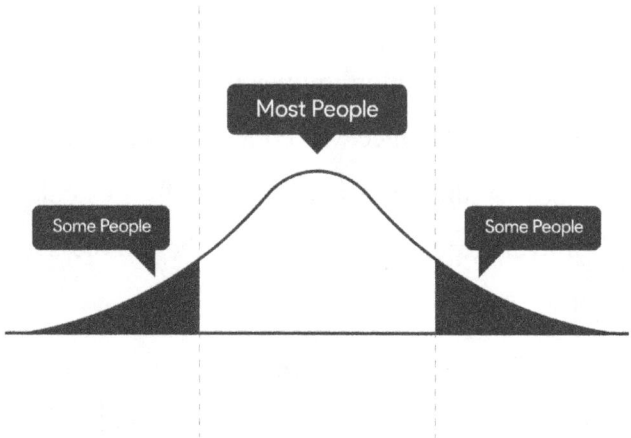

Picture 6: The Bell Curve

In probability theory, the normal distribution is a very common continuous probability distribution. Normal distributions play a crucial role in statistics. They are frequently applied to represent random variables with unknown distributions in the natural and social sciences.[xxiii]

Normal distribution operates on the basis of the Law of Large Numbers. The average of the result we get from a large number of experiments should be close to the anticipated value and will get closer the more experiments we conduct. The law of large numbers is important because it promises solid long-term outcomes for the averages of random events.

Take a roulette machine in a casino as an example. The casino may lose money in a single round, but based on a larger number of spins its earnings will tend towards the predicted percentage. The winning streaks will be balanced over time by the law of large numbers bringing the net gains and losses back to the predicted average.

Long tail distribution in the nonlinear world

In the realm of nonlinear systems the idea of normal distribution is no longer applicable. There is no such situation as normal. Power laws manifest in what we call a long tail distribution in nonlinear systems. These are not normal, rather extraordinary occurrences impossible in normal distribution but can happen in nonlinear systems. They are called Black Swans.

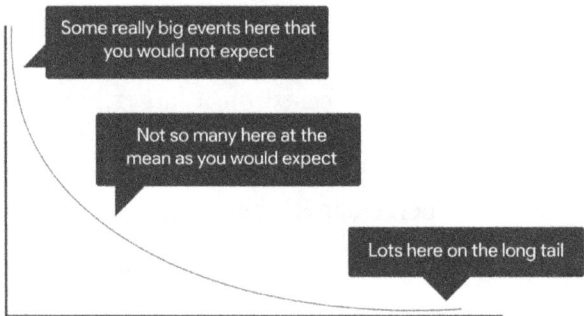

Picture 7: Long Tail Distribution

Who would have thought that on October 19th in 1987 the Dow Jones stock market's index would drop 22 percent in just one day? Its typical day-to-day fluctuation was less than one percent normally. Compared to that, 22 percent was a very not-normal, extraordinary change. We call this day Black Monday even today. In the realm of normal distribution such a deviation is unconceivable. But it happened nevertheless. How could people make sense of it? Only by using something else than normal distribution as an explanation tool.[xxiv]

Such kinds of events are explained by power law distributions where the graphic illustration has a long tail, as you can see on the picture. This is where the name "long tail" comes from.

These occurrences dramatically impact the system's behavior and can completely derail the system's established "average" or "normal" state. This is a tricky thing because if we just take any random sample to analyze a system's behavior, and there are no black swans in that sample, we'll get a nice average and normal state prediction. Just like the Dow Jones; if we only analyzed the sample of ten years' activity closing with October 18th, 1987, we would have got a stable trend of activity. But if we add just one more day, the 19th of October, 1987, that would have changed our map drastically.

Let's recall the example of people's foot size. We had a sample of men whose average foot size was 10. Now imagine that the man with the largest foot in the world walks in, the average size might change. Or take an even more dramatic example, we have a room full of people and we want to know their average income. If Jeff Bezos were one of the people in that room, the average would be radically misrepresentative. One billion dollars as average income. Sounds great, right?

In power law distributions, increasing sample values don't converge to a certain average but rather diverge from it. Exceptions apply, of course. Looking at power law distributions through the lens of probability, they present an exponential connection between the size of an occurrence and its frequency of happening.

There are so many examples in our world that follow power law distribution with long tail graphs. For example, the occurrence of earthquakes, stock market crashes (as we saw), presidents being shot. But there are less dramatic examples, as well.

Take professional photography. There is a pool of exceptional photographs created. Some of these photos were taken by trained professionals, but many by highly skilled amateurs named "everyone else" on the graph below.

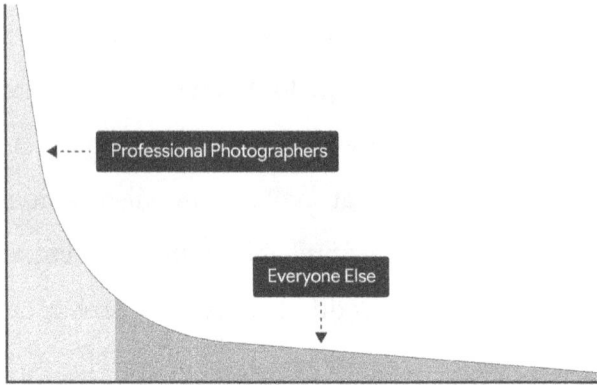

Picture 8: Long-Tail Distribution[xxv]

We can see that the best quality photos are largely made by professional photographers with great cameras but the trend is followed by a large number of amazing photos made by "everyone else." The professionals may produce the largest number of highest quality photos but cheerful individuals add more to the table through their unique viewpoint and personal touch.

The extreme events occurring in power law distributions remind us how nonlinearity is the product of interactions between elements and over time.

This power law makes extreme events much more likely and renders our traditional conception of average and normal no longer applicable. The law of normal distribution rests on reductionism; we take a random sample of elements and analyze them in isolation, having no correlation between them. Flipping a coin for instance once doesn't influence the second flip. In nonlinear systems elements are arranged in an atypical way; a big internet page like Google is big because of the network effect which is possible because people have chosen to use it. It seems random, but in reality it is not.

Chapter 5: Dynamic Systems Introduction

As we discussed already, systems are complex things having elements, interconnections and purpose or function. The interaction between the elements can't be explained by the individual elements alone. The interactions within a system create something more or less than the sum of the addition of each part. We can see this manifesting in biology, physics, human behavior and many other natural and artificially created phenomena. We understand by now what systems are. But what are dynamic systems?

Let's define dynamic. Simply put, it means change over time. Systems in most cases are not static. Even minor changes, activities between the elements can lead to a shift in the behavior of the system.

The term dynamic system originates from the language of mathematics. Dynamic systems theory works with mathematical equations therefore to model and calculate the changes over time in a system.

Dynamic systems theory is interdisciplinary, meaning it involves and mixes various theories to explain changes over time. These theories are chaos theory and catastrophe theory, among others.

We will discuss chaos theory in detail later in this book. Here I'd like to provide only a brief introduction. The behavior of chaotic systems seems random and unpredictable. Chaotic systems can be ecological, like the weather; social, like a blind date; or economic, like the stock market. Our ability to make predictions about these systems is limited. We could say chaos has low entropy as predictions are limited. (Entropy is the capacity to predict what comes next based on everything we knew from before about that specific subject.)

Thus in real life, chaotic system-related predictions are rather estimates or approximates. We can know approximately what the temperature will be today but we don't have complete information about every molecule in the atmosphere related to the weather. We don't follow every minor shift in these particles and thus we are partially blind to events that could develop in an exponential fashion.

Interestingly, catastrophe theory isn't the discipline of worldwide tragedy analysis. Catastrophe theory examines sudden changes in behavior. It has been able to explain how rapid, unexpected changes in behavior can originate from minute changes in the control factors. Take a stuck screw on your bicycle as an example. You're trying to unscrew it but it doesn't seem to move. As you increase the pressure added into your unscrewing, at some point, the screw will suddenly move and from that point onward you'll be able to remove it. As increasing pressure is applied, nothing happens, and the door remains stuck. The transition from stuck to unstuck is what we call a catastrophe.

Dynamic systems theory focuses on understanding the system as a whole. For this purpose it takes a step back and observes the full system from a distance rather than micro-analyzing the smallest components. For example, paying attention to the atomic structure of a plant won't answer your question of what is needed to create a healthy environment. There are many elements within a system that both constrain and create the so-called macroscopic emergent behavior which lays in the focal lens of dynamic systems theory.

We discussed at the beginning of this book that scientists preferred reductionism in the past, breaking the system down into individual parts and seeing how they influence each other. But as the researchers keep on breaking the system down to smaller and smaller fragments, the study becomes increasingly detached from the real, holistic behavior.

We now know that systems rarely behave in a linear fashion. System behavior can be surprisingly stable even if large changes happen within their elements. However, we can experience extremely dynamic behavior and changes by small shifts within the interactions of these elements.

Before getting deeper into the subject of dynamic systems and chaos let's go through some of the fundamental concepts and terms of dynamic systems theory.

Emergence and self-organization

Emergence is a basic presumption of dynamic systems hypothesis that proposes that the interaction of the elements of a system create an example of behavior that is new or not the same as that which existed earlier.

We experience emergence daily as emergent occurrences are visible, tangible and touchable. Take the weather as an example. A hail storm doesn't happen for no reason but rather is the result of the interaction between weather elements like temperature, humidity, etc.

Emergence happens as a consequence of the interaction of the system's elements and the idea which indicates how these parts interact is called self-organization. Thus we can say that self-organization is the room within which emergence happens.

There is no leading element that dictates how the other elements should interact. The interactions are led and influenced by feedback loops. Patnaik and Jostad summarized what the key conditions of self-organization within a system are:

"1) Interactions within systems must be *nonlinear*. Patterns emerge because of the different types and levels of interactions that can occur between the elements of the system.

2) A system must be *far from equilibrium*. This means that a system must be an open system and not a closed system. Although closed systems really only occur in a vacuum, the idea

is important. Systems must have a constant flow of energy in and out of the system. If this does not occur, then a system would find its equilibrium and essentially die. In social psychology, we often refer to this "energy" as information.

3) Systems exhibit *hysteresis*, which literally means "history matters." The future state of a system will depend on the past and present states.

4) *Circular causality* is present in systems. The pattern (emergence) is restricted by the behavior of the elements of the system, but the elements of the system are also restricted by the global patterns.

5) *Fluctuations or perturbations* are constantly trying to move the system, which will provide an understanding of the stability of various

phases. These perturbations should not be misinterpreted as error.

6) The *slaving principle* acts as a selection mechanism for the interaction of the parts of the system (Kelso, 1995). That is, once the order parameter (see below) is fully developed and rather stable, it "enslaves" the parts of the system into specified interactions."[xxvi]

Let's see an exemplary case of emergence through self-organization called the Raleigh-Benard instability.

The Raleigh-Benard instability

The Raleigh-Benard instability shows the emergent behavior of a liquid heated from the bottom.

First think about oil inside of a skillet which is not heated. When the surrounding temperatures of the oil are the same, the oil molecules move around in a random fashion. But when we start heating up the oil, the movement pattern of its molecules changes. The molecules that reside on the bottom of the skillet, due to heat, will rise to the top of the oil's surface. The denser and colder molecules will fall to the bottom of the pan. At a specific temperature gradient (contrast in temperature between the base and top of the skillet) a pattern emerges inside the oil that is created by the motion of the molecules which in this case are the elements of the system. The molecules interchange their position in a rotated manner, as you can see in Picture 9.

Picture 9: The Raleigh-Benard Instability

The warmest molecules rise to the top of the oil where they get distanced from heat so they eventually cool down and sink back to bottom from where the previously colder molecules rise, now them being the warmest. If the temperature stays stable, we gain a self-organized pattern from a state of disorder. But if the temperature keeps rising, the oil becomes unstable again, falling back to a state of disorder or chaos.[xxvii]

Sensitivity to initial conditions

Dynamic systems are sensitive to initial conditions. We discussed this briefly when we talked about nonlinear systems. Sensitivity to initial conditions means a small initial difference in the starting conditions that over time can produce great differences. Chaos theory relies on this fundamental idea. Logistic map equations present this idea fairly well. This is the equation…

$$x_{t+1} = Rx_t(1-x_t)$$

… where x_t is the present value of x and x_{t+1} is x's next value. R is a constant. This equation tries to illustrate that even minor changes in the value or R (the constant) can bring dramatic, long-term changes. Just substitute R, x, and t with numbers and calculate what happens if you repeat the process, say, ten times.[xxviii]

Let's see how a logistic map looks like as a diagram. Take a look at Picture 10:

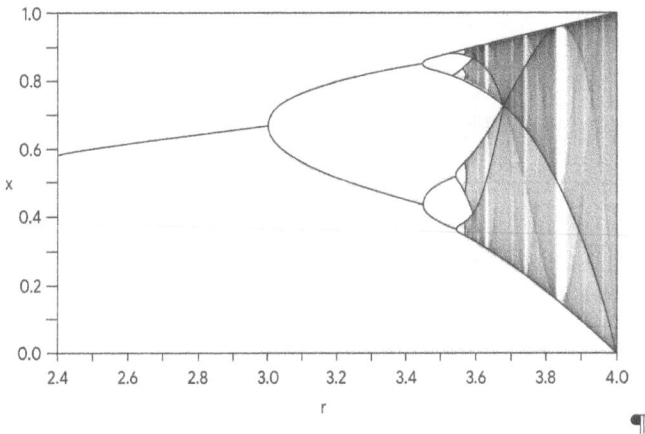

Picture 10: Bifurcation Diagram

This particular diagram shows that below about 3.1 the R, constant is composed by fixed points (the straight line). The first bifurcation is triggered by a minor change in initial conditions that happens after 3.1. Now, as a consequence of the small change we have two values for our R constant. Around 3.5 another bifurcation happens and now we have four different values. The higher the value of the R constant, the bifurcation repeats itself again and again, doubling each time the number of values. We call this phenomenon period doubling. The periods continue to double until they arrive to a chaotic state that you can see on the diagram happening around 3.6. The bifurcations are the indicators of a phase transition present in the system.[xxix]

Let's define what phase transitions are. In general, phase transition as a term is used to present transitions between solid, liquid and gaseous states of matter. "During a phase transition of a given medium, certain properties of the medium change, often discontinuously, as a result of the change of external conditions, such as temperature, pressure, or others. For example, a liquid may become gas upon heating to the boiling point, resulting in an abrupt change in volume."[xxx]

A phase space of a dynamic system is the gathering of every single condition of the system being referred to. The estimation of the outside conditions at which the change happens is named phase transition.

Dynamic systems stability

Dynamic systems want to accomplish and keep up a stable state. At the point when a system is knocked out of balance, it embraces certain patterns which endeavor to accomplish local stability. This is achieved with the utilization of order parameters and control parameters. Stability is a protection from disruptions, which are minor changes or flusters in the system.

There are a few different paths to calculate the stability of a system. Stability is recorded by the statistical probability that the system will be in a specific state instead of other possible states. Stability can also be estimated depending by the way it reacts to disturbances. If in the event of a minor disturbance the system gets pushed away from its stable state, over time the system will swing back to its starting position. Stability is connected to the system's reaction to normal fluctuations inside the system.

In general systems theory, a system is stable if its various elements are not changing over time.

Order parameters

Order parameters help the repetition and differentiation of a coordinated pattern of movement from other patterns within a system. Order parameters are the phenomena in which we are trying to gain insight about system patterns.

Phase transitions happen when order parameters change as a function of another parameter. Take temperature as an example. An order parameter is showing the level of order over the limits in a phase transition system. In case of fluids or gas, density difference is the order parameter.[xxxi]

Control parameters

Control parameters change the stability of states by acting as a catalyst for restructuring behavior. We don't talk about control in the traditional sense when we mention control parameters but rather about a state which is sensitive to the combined behavior of the system. When this control parameter is triggered, it moves the system through collective states leading to phase shifts that threaten the stability of the present attractor.

For example, in Raleigh-Benard instability's case, heat acted as a control parameter being an outside variable pushing the system into divergent behaviors.[xxxii]

State space

State space illustrates the probable states of the collective variable. If we analyze a simple object as a pendulum, we can describe its state in a two-dimensional state space. The two factors we need to consider are position and velocity. The pendulum swings there and back, having a path through the state space. We can track there the changes of position and velocity.

Let's take a look at a three-dimensional state space now. Topology is a good example; an elevation map of a territory for instance.

Picture 11: Three-Dimensional Elevation Map

Picture 11 is a topographic map near the area of the University of Utah. It measures the distance and direction on the X and Y axis on one hand, and elevation through the lines on the other hand. It stands for a three dimensional space. Imagine someone was placed onto a point of this geographical territory. Chances are high that our person will start walking towards the path of least resistance; somewhere downhill or on a flat surface.

Systems work the same way. That is, based on where the present state of the system is, a system topology can show us the most probable future dynamics of the system.

Continuous and discrete system states

As we said, a state space of a dynamic system is a two- or possibly three-dimensional graph. Here we can present all possible states of the system with each possible state of the system corresponding to one unique point (Set Point, see later) in the state space. We can model the change in a system state in two ways. These are called continuous and discrete states.

Continuous

A system is in a continuous state space when the time intervals between our measurements are negligibly small. This makes the state look like one long continuum. We do the measurement through the language of calculus and differential equations. These are a good way to measure when we're dealing with an abundance of elements which each hold a lot of information. These measurements can become too complex extremely swiftly.

An example of a continuous system is one in which the state variables change continuously over time, like the amount of water flow over a dam.[xxxiii]

Discrete

When we measure state space as discrete we have a discernible time interval between each measurement. In a discrete system the state variable changes only at a discrete set of points in time. For example, ants leaving the colony one hour after sunrise, two hours after noon, etc.

We use iterative maps to measure this. Iterative maps or functions grant access to less information but they are simpler and more fitting to deal with many elements where feedback is important. Iterative maps are crucial to study and understand nonlinear systems. They allow us to take the output of the previous state of the system and feed it back into the next iteration. Iterative maps are great thus to capture the feedback loops. Iterative maps also have almost the same type of attractors (read about them later) and bifurcations as dynamic systems.

Patterns of behavior in the system

Self-organizing systems usually display a limited number of patterns even though in theory they could display quite many. A dynamic system can usually be described by one or a few differential equations, which specify the system's behavior over a short time.

Attractors[xxxiv]

Attractors are physical properties that drive the system to evolve around them. The starting conditions of the system don't matter, the attractor will draw the system toward its state space. System values that get close enough to the attractor stay close even if slight disturbances happen.

If you imagine a graph that shows the changes in a system, the attractor's influence would look like a negative slope. The steeper the slope the faster the system's points are drawn towards the attractor. In real life, imagine how two people who feel attracted to each other would act. The bigger the attraction, the closer they'd like to be.

We can enumerate more real life examples of attractors such as the use of addictive substances which create addiction in our body's cycles by their physiological influence. When we satisfy our craving for the substance the urge of attraction subsides but continuously comes back to us in a periodic and predictable way. Unless we are able to break free from the substance, the craving will return.

A basin of attraction shows every point within our state space that moves a system toward an attractor. For example Earth's gravitation field is a basin of attraction. If we put something large enough into the gravitational field it will be attracted into the platen's orbit regardless of its starting condition.

Strange attractors

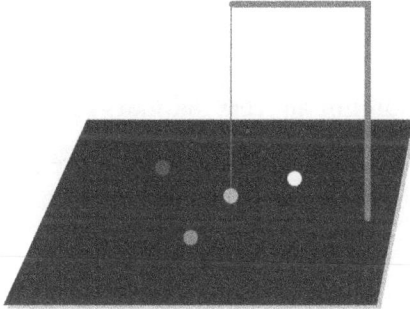

Picture 12: Magnetic Pendulum

Imagine a pendulum with a magnet at the end of it, as shown in Picture 12. The pendulum moves independently between the three magnetic points, shown as white, light grey and grey. Each point elicits equal attraction to the magnetic pendulum. As the magnetic pendulum circles around the three points, it will eventually come to stop over one of them.

However, where the pendulum will end to stop depends very much upon the starting point. If you start the pendulum over one of the points, it will stop over that magnet. But if you release the pendulum in a random fashion, it is very hard to predict where it will stop. We can start the pendulum in very similar, close positions but we may end up with three different outcomes.

What we can do to describe the strange attractor for this system is to map out all the outcomes for all the possible starting positions of the pendulum. This can be done with a computer simulation. Even though we are talking about a minor mechanic system like a magnetic pendulum, the calculus of the attractors is extremely complex. Because the setup has three symmetrical points, the attractor is also symmetrical. The attractor is very sensitive to initial conditions; minor changes in the starting position can result in great differences in the outcome.[xxxv]

Repellers

Repellers act as boundaries between attractors and they have a positive slope. A less steep slope suggests that the points are moving away slower. Taking the same example, two people who are repelled by each other will try to keep their distance; the greater the state of being repelled the more distance they'd like to have.

Set points

A set point is where we choose one specific point within a system and describe all behavior in a system in relation to this point. Attractors try to get closer to the set point and repellers try to get away from this point.

Set points show no behavior change. When there are oscillations in the system, this is the point the system oscillates around.[xxxvi]

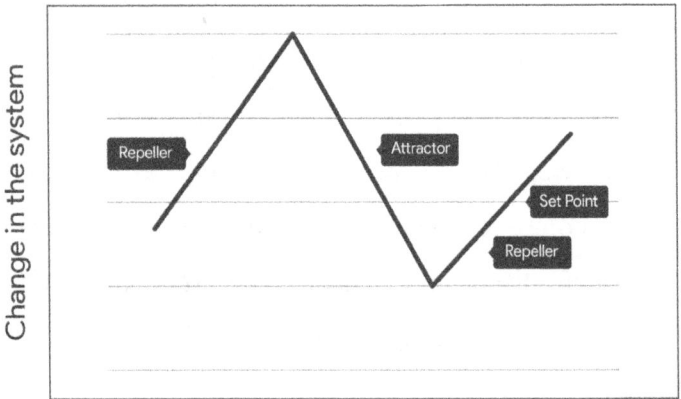

Picture 13: Attractors, Repellers, Set Points

Periodicity

Periodicity measures how a system returns regularly to the same or similar state in the case of a pattern repeating itself infinitely. Periodic behavior recurs at regular intervals, for example every hour or every year. A period is the amount of time it takes to complete one cycle. A dynamic system having a stable periodic behavior can be called an oscillator. An oscillation is a repetitive combination of conditions between two or more states. Take a look at Picture 14. The S-shaped condition is a period and the constant up and down motion is the phenomenon of oscillation.

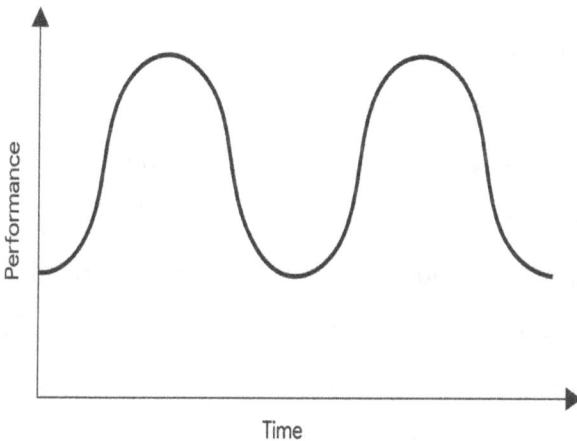

Picture 14: Periodicity and Oscillation

Limit cycles can present a lot of real world oscillatory behaviors over time. A pendulum is a limit cycle as, if it is not affected by friction, it will repeat its oscillations forever. Stable limit cycles have self-sustained oscillations describing a harmonious periodic behavior of the system. After any small disturbance in the closed trajectory the system will return to the limit cycle.

The state a dynamic system gets to after an infinite amount of time and repetition is called a limit set. We can use limit sets to understand the long-term behavior of the system. Attractors, for instance, are limit sets. However, not all limit sets are attractors. Imagine a pendulum losing speed. Before, X was the pendulum's minimum height and Y was the maximum height. X therefore is a limit set because the trajectories move towards it, but not Y. Also, since the pendulum is losing speed, point X is the attractor.[xxxvii]

To give a more concrete example of attractors in real life, think about the motion of the planets around the sun. This is a periodic motion; a very predictable one. We can predict far in the future and way back to the past when eclipses happened, for instance.

In these systems small disturbances are often rectified. They do not alter the system's trajectory very much in the long run. Opposite to the state of stable equilibrium, here the system cycles around some equilibrium.

All dynamic systems require some input of energy to drive them. In physics these input energies are referred to as dissipating systems. Systems repeatedly dissipate the energy being inputted to them in form of movement or change. A system and its trajectory being bound to its energy source will follow some periodic motion around, towards or away from this source. In the case of the planets' orbit, they follow a periodic motion because of the gravitational force the sun forces on them. If the sun would not elicit this driving force, the motion would cease to exist.

Another real life example of periodic behavior is the human body which needs food, rest and other things on a regularly repeating basis. We eat and then burn the calories through some activity and then eat more and consume it again. We are bound to cycle through these sets of states.

Behind the complexity of how the real world works is the fact that a system's dynamics can be the result of many, various interacting forces that have more than one attractor states. Systems can shift between these different attractors over time.

Take a double pendulum as an example. A single pendulum follows the periodic and deterministic motion features of linear systems having one equilibrium. Let's put a joint in the middle of the arm of the single pendulum. Now it has two arms instead of one. Because of this change, the dynamic state of the system will be a result of these two arms interacting over time. Thus we get a nonlinear dynamic system out of a linear one.

In our previous example we analyzed the dynamics of a planet orbiting around the sun. Here we have a single equilibrium and attractor. But what if we added another planet into this equation? We now have two points of equilibria creating a nonlinear dynamic system. Why? Because now planet A is under the influence of two different gravitational fields of attraction. There was only one basin of attraction so it was not important where the system started out, it would only gravitate towards the only equilibrium point. But now we have multiple interacting basins of attraction. Think about the magnetic pendulum example. Now the starting point of planet A matters as small changes in its initial state can bring forth quite different long-term trajectories.

This is what we call chaos.

Chapter 6: Chaos Theory

Chaos theory was developed by mathematicians to describe the behavior of "dynamic systems," or systems that entail the location of a point in geometric space at given times. The dynamic systems that chaos theory deals with specifically are those that are heavily influenced by their initial conditions, or the starting point at which the system begins its motion. A good way to think about this is thinking of the "butterfly effect," or the idea that a butterfly flapping its wings in Indonesia can cause a hurricane in Florida. The small initial condition—the butterfly flapping its wings at a certain time and rate—has a massive effect on the larger dynamic system of the weather (which is dynamic because it entails wind and water moving around different points at specific times in the atmosphere). Systems to which we apply chaos theory tend to have extreme differences in outcomes depending on the initial conditions—if we think of the butterfly, for example, the consequences could be a hurricane somewhere, or a tornado, or nothing at all depending on when the butterfly flaps its wings.[xxxviii]

Unsurprisingly, chaos theory lends itself to applications in climate studies very well; it also helps in predicting and forecasting traffic patterns, which depend on decisions made by drivers that have a wide variety of outcomes. In fact, many different fields use chaos theory, especially those dealing with outcomes of human behavior such as sociology, economics and political science. What these uses all have in common is that they are interested in mathematical chaos, rather than what the average person would think of as "chaos." While most people probably conceive chaos as general disorder and mayhem, mathematicians think of chaos as something that is so dependent on the initial condition it is nigh impossible to predict. For example, if you set the initial condition to the number 1 and the final condition is 10, the mathematical equation that provides that result will yield a different result if you start with 1.01 or 1.2. If this result is pretty close to 10 (basically, a difference of less than 1), then the system doesn't have any chaos. But if changing the initial condition to 1.2 yields a result of 14, then the system has a high level of chaos because it's not very predictable.[xxxix]

Chaos theory generally deals with simple deterministic systems, for example a scenario where one human's decision would affect an outcome in different ways, as opposed to the effect of many people's decisions. Complex systems with many different initial conditions would naturally produce many different outcomes, and are so difficult to predict that chaos theory cannot be used to deal with them. Chaotic behavior of simpler systems is also more interesting, because it is so unexpected due to the relative "un-randomness" of their nature.

Chaotic systems are largely determined by chance, but it is not random. Indeed, the initial conditions are highly deterministic in how the system plays out. However, because they are deterministic in this way, it is very difficult for humans to predict the outcomes. This might sound counterintuitive, but it is actually proven in mathematical and practical contexts.

This unpredictability stems from the interactions between the components of the system after the initial conditions are set. This unpredictability is also part of complex systems, of course, but those are not so easily described mathematically. However, it is fairly simple to put them into practical terms. Think, for example, of your commute to work. If you ride the subway, you walk to that station expecting to ride it. However, the subway is having work done on it, so it's closed. So you can choose between walking or riding the bus. Either of these scenarios will put you around a different set of people, each with their own conditions for being where they are, and therefore entail a chain reaction of different interactions. In complex systems, initial conditions can be especially impactful during certain points while they develop, such as the subway closing or a certain person boarding the bus. Each decision you make marks a point where the system encounters an imbalance, which can lead to drastically different outcomes depending on which choice wins out. Imagine, for example, that you will encounter a scrial killer on the street while you walk, or the love of your life if you ride the bus—this is an unusually extreme contrast, but it also demonstrates how differently a life can go based on the path chosen at a seemingly nonsignificant juncture.

The development of chaos theory in mathematics took a long time, and is interesting in itself. The French mathematicians Henri Poincaré and Jacques Hadamard laid the groundwork for the development of chaos theory in the late nineteenth century. Poincaré's work on the three-body problem, which is a physics problem that entails solving for the motion of three point masses from their initial positions and velocities. He introduced the issue of chaotic orbits, which were not periodic but neither approached a fixed point nor infinity. This meant their motion did not conform to existing mathematical models.[xl]

Jacques Hadamard, meanwhile, was the inventor of the Hadamard's billiards mathematical scenario, wherein a free particle has frictionless, chaotic motion over a surface that curves away from its geometric plane in two different directions; think of this kind of surface as a saddle shape. Hadamard's billiards is an important mathematical achievement because it demonstrates that all physical trajectories are chaotic; that is to say, they derive instability from the vast changes initial conditions can wreak.[xli]

As is fairly easy to surmise from these developments, chaos theory arose mainly from research in physics. The development of chaos theory in earnest began with the study of ergodic theory—the behavior of dynamic systems over time—and nonlinear differential equations, which describe change over time that does not progress at a constant rate. Studies in this area tended to focus on physics problems like astronomy, radio engineering and the three-body problem, where they had observed anomalies in fluid motion and oscillation, or wanted to investigate the possibility of such anomalies as was the case in astronomy. Chaos theory was not acknowledged as a field in its own right until the latter half of the twentieth century, when mathematicians and scientists realized that there were too many quirks in their data to conform to a linear theory of systems. Electronic computers also accelerated the development of chaos theory because of their ability to repeat calculations of formulas, which was needed for the study of chaos theory but too time-consuming to be practical when done by hand. The development of computer graphics further aided in study by providing quickly generated visual depictions of problems.[xlii]

One of the first prominent examples of the usefulness of computers to chaos theory is the case of Edward Lorenz, a meteorologist and mathematician. Lorenz was running a weather simulation on an early digital computer in 1961. When he repeated the simulation to see a certain set of data again, he restarted it in the middle instead of running it from the beginning. This led to the generation of an entirely different set of weather data than the one he had originally gotten. Lorenz found that a small difference in the initial conditions due to decimal rounding had led to the entirely different outcome. This phenomenon later became known as Lorenz Attractors in forecasting; it also demonstrated the near impossibility of long-term weather modeling. As a note—this is different than climate modeling, because weather describes short-term localized patterns while climate is a much more long-term set of data designed to show larger trends, so Lorenz's difficulties do not affect the effectiveness and credibility of climate science.[xliii]

Another key pioneer of chaos theory was Benoit Mandelbrot, originator of the Mandelbrot set and fractal geometry. Mandelbrot was a mathematician who coined the terms "Noah effect" and "Joseph effect" for sudden changes not consistent with existing patterns in pricing. His idea of fractals originated with the idea that measuring instruments of different sizes will change the length of a coastline; using this and the example that a ball of twine appears smaller the further away the observer is, he proposed that dimensions are fractional and dependent on the observer's position. If an object's dimensional traits are the same on different scales, it is a fractal. To put this in simpler terms, think of a fern. Although its leaves get smaller and smaller close to the top, their size and pattern are the same; this is the rough concept of a fractal. Many fractals occur in nature, including frost crystals, mold, Romanesco broccoli and (ideally—trees are highly regular in reality) trees. Fractals' "branching" nature is important in chaos theory, which can be conceptualized as the generation of different "branches" from a slight change in initial conditions.[xliv,xlv]

Through the 1970s, chaos theory developed as its own subfield and began to generate its own conferences. Its applications also became more diverse through the 1970s, 80s and 90s, branching out into medicine, biological sciences, natural disaster prediction, evolution, financial market fluctuations and even warfare. Chaos theory is also an important component of cryptography and code breaking; much current encryption technology, including pseudo-random number generators and digital watermarks, makes use of chaotic mathematical principles. Chaos theory has also been used to improve robot and artificial intelligence prediction of future outcomes and problems.[xlvi]

Chapter 7: The Butterfly Effect

Now that we have a more comprehensive understanding of chaos theory, let's explore the Butterfly Effect in more depth. The Scientific Method relies on isolating a system to study it; this is the entire principle behind the concept of the scientific experiment. However, in reality this is impossible. Every single particle, animal, vegetable and mineral has an effect on everything else that exists. For example, every bit of matter, even a single electron, exerts gravitational force on other pieces of matter in the universe. And this is only one way in which every single thing affects everything else. Although experimentation does its best to test only the variables it wants to, in reality this is a goal that is unattainable given its scope.

Establishing causality is much more difficult than people would like to believe. Saying that one thing causes another means removing all the other possible variables that could have an effect, which is impossible. However, the scientific method is important because it helps distinguish causes and effects that are "significant," as opposed to "insignificant." This is the entire idea around the concept of "significant figures," which many people learn in high school science. The idea is to refine one's data to a decimal point that will show the line where results become significant, and where they don't. Statistical analyses also traffic in significance; for something to be statistically significant, it has to be statistically uncommon for it to happen *unless* there is a relationship between the variables studied.

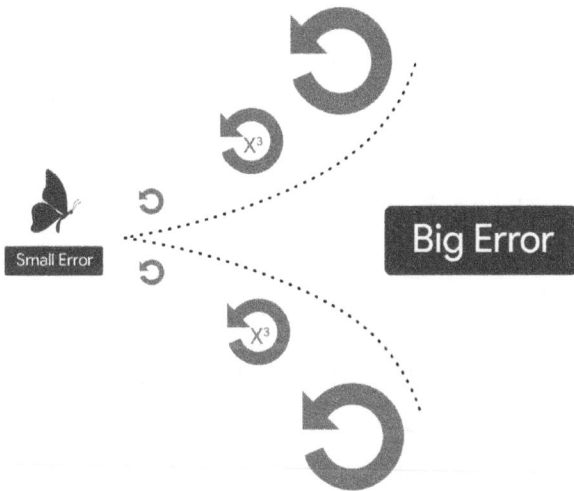

Picture 15: The Butterfly Effect

Another example of the power of significance is the phenomenon we've already discussed: gravity. If one were conducting a simple physics experiment to test the acceleration of a ball and a feather when dropped in a vacuum, a black hole in another galaxy does not need to be taken into account, even though it exerts an extremely strong gravitational force. This is because the black hole is so far away that its gravitational force on Earth is so small as to be effectively nonexistent (think of how close .000000000000001 is to zero, compared to .01). The human brain processes these tiny variables as not existing all the time; they are simply beyond the scale of what we would count as a perceptible effect. Human brains are generally wired to protect us from the anxiety that telling the waiter to keep the change could crash the global economy, or that getting dirt in a scrape will lead to a septic infection (however, this is not to say that all brains are like this). If all the variables in a system were equally important, the world would be so chaotic as to be impossible to live in. However, because many are so small, we can take comfort that many systems in our lives—such as generous tipping—are quite linear.[xlvii]

However, the systems to which we apply chaos theory are nonlinear, which means that differences in outcome can grow exponentially through feedback loops. This means that any small effects of changes in initial conditions can be amplified the longer the system runs, because they can cause more effects and divergences the longer they exist. This is exactly what happened to Lorenz— small differences in his data in an initial iteration were highly amplified the more he repeated his computer's data processing. Although Lorenz's discovery was highly technical, we can also think of it as the butterfly effect—namely because Lorenz named it such. He chose this metaphor not only because it related to his profession as a meteorologist, but also because it is an easy concept to understand.

Lorenz realized that the butterfly flapping its wings was like the slight discrepancy in his data; it seemed small and like it had no effect initially, but the system's subsequent functioning was deeply affected through a chain of feedback loops that the initial condition caused. Furthermore, these loops and effects increased exponentially as more and more were added on by the system. It's not that the system is impossible to predict, it is just extremely difficult. Although scientists once disdained the idea that the butterfly effect could be a reality, Lorenz and others' work explained the quirks of dynamic systems that people had long struggled with, and they continue to be a cornerstone of the scientific and mathematical communities today.[xlviii]

The butterfly effect is, unsurprisingly, still most prevalent in weather models, in which chaos is a crucial component of modern prediction methods. However, a butterfly flapping its wings would not be a literal concern of meteorologists. The butterfly effect would take too long to develop in a system to be useful in weather forecasting, which is generally done in the short term (which is evident in the standard ten-day forecast model found in the United States). Thus, although the concept it describes is extremely important, Lorenz's metaphor in a literal sense is not.

Scientists are also currently investigating the presence of the butterfly effect in quantum mechanics. Quantum mechanics is a relatively new field of physics that has seen rapid development in the computer age. Several scientists have claimed that there is evidence to expect the butterfly effect to appear in quantum systems which have slightly different Hamiltonians—Hamiltonians are the sum of kinetic and potential energies of all the particles in the system. Thus small differences in initial system energy could have huge effects in a quantum system. Some scientists believe that they have found the butterfly effect in the form of fidelity decay, which is the rate of divergence of identical initial conditions when subjected to different systems (think of this as the inverse of the chaos theory principle). The application of chaos theory to quantum systems is known as quantum chaos.[xlix]

"The butterfly effect" has become a common colloquial term in our society. However, the colloquial sense is not necessarily compatible with the scientific sense. Many popular culture usages seem to take the "butterfly effect" as the idea that every event that occurs on earth has a small causal event that indirectly caused it. This feeds into humans' desire for the universe to have order, for everything to have happened for a reason no matter how small, but it is not the proper scientific root of the term. Instead, this conception is more in line with Schrödinger's theory that each event is a point of divergence for parallel universes.[1]

To think about the butterfly effect and chaos theory in a more concrete sense, we should undertake some real world thought experiments and see how they turn out. Say, for example, a professional tennis player was serving a tennis ball. If they serve the tennis ball repeatedly, it will probably end up in roughly the same area of the box; professional serves are generally consistent enough to predict linearly. Now, imagine that the tennis player serves two balls at once; suddenly, it is extremely difficult to predict where *either* of the balls will end up. The system of the serve is sensitive enough to the initial condition of the number of balls that any alteration will cause chaos! What other systems could be the same way, where changing one initial condition will vastly alter the outcome of the system, generating high levels of chaos? How would we go about proving that system is chaotic? If it feels comfortable to stick to a weather application, think about weather forecasts and their range. Are predictions more accurate when they're made closer to the time they're forecasted for? At what rate do they become more accurate? And, most importantly, if they do, *why* do they become more accurate?

Here's a hint—carefully think back to Lorenz's experiment, and the circumstances under which he discovered the butterfly effect.

Chapter 8: Phase Transitions and Bifurcation

In the previous chapters, we discussed the phenomenon of feedback loops. But how, exactly, do these loops appear and function in dynamic systems, especially chaotic ones? First, it is important to get a handle on the different types of feedback loops.

Negative-balancing feedback loops limit change; this means that they help keep a system more stable, because they counteract factors that would make it change.

Positive-reinforcing feedback loops, meanwhile, accelerate change in a system in a way that is so rapid as to be unstable. Positive feedback loops are also unsustainable because they cannot exist indefinitely in an environment—all environments place checks on a system's growth eventually. This means that the exponential growth we'd be likely to see in a chaotic system that has positive feedback loops will eventually stop, because of the physical limits imposed on it.

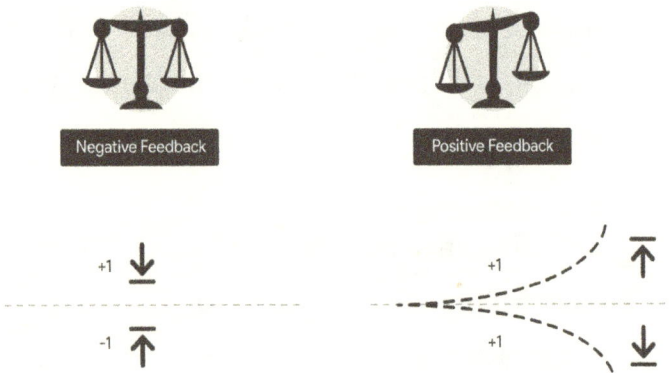

Picture 16: Negative and Positive Feedback Mechanisms

Most of our universe is held in a relatively stable condition by negative feedback loops. For example, the population of insects has not increased dramatically because there are still plenty of birds and other predators to eat them. In fact, the population is decreasing, likely due to climate change, but that is an entirely different feedback loop. We can also think of the phenomenon of deer overpopulation in suburban and urban areas in the United States, which is driven by both natural predator loss and a decrease in the length and popularity of deer hunting season. Both of these examples are emblematic of positive feedback development, which can lead to what scientists call phase transitions.

Phase transitions[li]

Phase transitions are when a small change to a quantitative input (we can think of this as the initial condition) results in a qualitative change in the system. These are easiest to think about in terms of high school chemistry; liquids and gases are the same substance, but their qualities are what differentiate them. Thus the change from water to mist is a phase change, because they are described in completely different qualitative terms even though they are made of the same matter.

Beverage fermentation is similar; brewing kombucha, for example, entails changing the conditions of tea by adding yeast to ferment it. The qualitative aspects of the drink are different depending on how long the fermentation occurs, or the number of bacteria, just to name a few parameters. Therefore, changing one of these quantitative parameters can vastly alter whether you have tea, kombucha or something so fermented as to be undrinkable. Tea and kombucha are made from the same substance, but these quantitative variables alter their qualities to the point that they are two separate drinks. We can think of this as "regime change" in the most literal possible sense—the rules and parameters that govern the substance no longer apply after the quantitative change is made.

Bifurcation[lii]

Bifurcation theory is similar to phase transitions in concept; it also deals with the effects of quantitative change, but instead of qualitative effects it studies environmental effects and the subsequent change in something's attractor state. The attractor state is the set of conditions something encounters based on its current environment.

This change in attractor state is a bifurcation, or a branching. Think of this as a choose-your-own-adventure sort of idea; if one path is chosen, something encounters one set of circumstances, and if a different path is chosen, there will be completely different circumstances. These circumstances are the "basin of attraction" because they also guide future events, and can provide their own jumping-off points. Each event trajectory therefore comes from a specific set of attractors. Bifurcation entails "regime change" because there is a new set of circumstances in equilibrium.

Think, for example, if you majored in dance in college at a conservatory. During your studies, your basin of attraction has been pretty wide; there are many different types of dance in the course of study, but you haven't taken totally different subjects like engineering, which is not in your basin of attraction. When you graduate, you are given the options of either joining a classical ballet or modern dance company. This represents a bifurcation in that each option has its own set of attractors, because it requires you to specialize in a specific genre of dance while your college studies did not. A small event, such as a mean comment from someone in your class who is joining the ballet company, could have a large impact in determining your trajectory into the next basin of attraction, which in turn will determine the new bifurcation points and basins of attraction you encounter later in life.

There are two types of bifurcations: local bifurcations and global bifurcations. Local bifurcations are tracked through local stability properties of system equilibria, where they can cause changes. This means that their effects are generally seen within a single part of the system itself and are fairly easy to spot. Global bifurcations, meanwhile, occur when larger parts of the system collide or otherwise disrupt the system's equilibria. This means that they cannot be spotted by tracking changes in the equilibria's stability, and instead must be found through more holistic analyses that spot changes in overall topology. This is where chaos theory, when applied to attractors to define "chaos attractors," becomes useful.

All this is to say that larger changes in systems equilibria tend to have more chaos because they affect more parts—think of the Titanic, where the watertight compartments were designed to prevent sinking if one or two flooded because they could be shut off from the rest of the system, but the flooding of nearly all the compartments proved catastrophic for the system of the ship.

Systems with exponential growth, otherwise known as nonlinear systems, tend to have many feedback loops and phase transitions that drive this growth. As a result they have a different development pattern than linear systems, which is called punctuated equilibrium. Punctuated equilibrium occurs

when positive and negative feedback loops
become out of balance. Negative feedback
loops hold positive feedback loops in check in
a linear system, providing the system with
equilibrium. This equilibrium means there is a
stable basin of attraction, and no phase
transitions. Punctuated equilibrium happens
when positive feedback loops overpower
negative ones, moving the system out of
equilibrium and into a phase transition. This
means it also moves into a new attractor state.

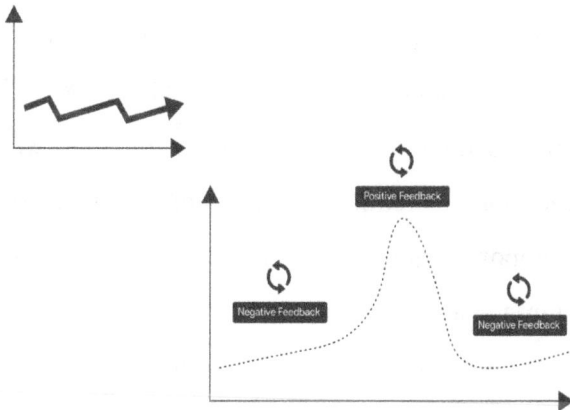

Picture 17: Punctuated Equilibrium

Perhaps one of the easiest to understand examples of punctuated equilibrium is the global financial crisis of 2008. Through the 1980s to the 2000s, the global financial system experienced relatively stable growth (at least, stable enough to satisfy an economist and not send the world into a massive multiyear recession) because its negative feedback loops in the markets kept runaway positive ones in check. However, deregulation and shady mortgage security practices destabilized the global economy's equilibrium to the point of the real estate crash of 2007, followed by the global financial meltdown of 2008. This represented a phase change in the world economy, creating a totally different basin of economic attraction. Many countries, such as Greece and Spain, have still not recovered and continue to live under austerity measures and economic stagnation; this is their new equilibrium.[liii]

Ecosystem collapse is another example of punctuated equilibrium, this time through species extinction and global climate change. This, again, creates an entirely new basin of attraction and therefore a new equilibrium in the ecosystem. Therefore, we can think of punctuated equilibrium as short periods of vast systematic change, between which there is little change or instability.

Chapter 9: Fractals

We've mentioned fractals already in this book, but it is important to have a deeper understanding of them. Fractals are essentially the repetition of the same pattern, but on different scales; therefore, if one zooms in on a fractal image, one will see what looks like the same image no matter how far they zoom in. Fractals also have to be detailed patterns. Famous examples of fractals include the fractal tree, the Sierpinski triangle and the Mandelbrot set. Visual representatives of all these fractals are available online, and it is very interesting to carefully view them and see the pattern.[liv]

Although the mathematician Gottfried Leibniz vaguely mulled over the possibility of fractals' existence in the seventeenth century in his writings on recursive self-similarity, calling fractals "fractional exponents," serious developments in fractals did not emerge until the nineteenth century. In 1872 the mathematician Karl Weierstrass demonstrated a graph of a function that was continuous but never differential to the Royal Prussian Academy of Sciences. In 1883 Georg Cantor developed Cantor sets, mathematical sets of points that were at the time categorized as merely unusual, but are now agreed to be fractals. Felix Klein and Henri Poincaré also made important contributions. In 1904, Niels Helge von Koch published a paper with a geometric definition meant to improve on Poincaré and Weierstrass's work, along with the Koch snowflake, one of the first proper fractal images. The Koch snowflake starts from an equilateral triangle, and recursively inserts more equilateral triangles (through two lines, coming to an equilateral angle, in the middle third of each line segment). In 1915 Wacław

Sierpiński created the Sierpinski triangle, and in 1918 Pierre Fatou and Gaston Julia developed more ideas about the roles of attractors and repellers in fractals. Soon afterward, Felix Hausdorff expanded the definition of "dimension" to allow for non-integer dimensions, providing more dimensional options for fractals. In 1938, Paul Lévy published a paper that included a new fractal curve called the Lévy C curve.[lv]

However, the most significant developments in work on fractals came from Benoit Mandelbrot in the 1960s and 70s, who was greatly aided by the development of computer graphics (earlier mathematicians had drawn fractals by hand). Mandelbrot indeed invented the term "fractal" in 1975, and published it in a paper that included several vivid computer graphics. These images were generated using recursive algorithms, or algorithms that repeat to infinity; recursive imaging and coding is now a basic skill taught in many computer science classes. However, there is also software that can generate fractals itself—without hand-coding—that was presented in 1980 by Loren Carpenter at the SIGGRAPH conference on computer graphics. The software, which he used to render fractal landscapes, was so impressive that Lucasfilm hired him for its Computer Division—later Pixar Studios—immediately after the conference. He remains at Pixar as its chief scientist, and has won several special Academy Awards (yes, Oscars) for his achievements. Thanks to Carpenter's work, it is therefore now easier than ever to generate and study fractals.[lvi]

To understand fractals, it is important to first understand the concept of dimensions. You have probably heard of the first, second and third dimensions, which can be represented by a line, a square and a cube, respectively. Fractals, however, are unique because although they are lines, they have fractional dimensions, and therefore bridge the space between the first and second dimensions. Coastlines, which Mandelbrot proved as fractal as previously mentioned, have between one and two dimensions, while a mountain, which is also geometrically fractal on its surface, will be somewhere between the second and third dimensions. As fractals have higher dimensions, it is more likely that any region of space one chooses will have a part of that fractal inside of it.

While chaos theory describes the wild, unpredictable parts of nonlinear systems, fractal geometry represents the parts that can be solidly organized into mathematical forms. That is why fractal images and naturally occurring fractals often seem so visually pleasing; they represent order and symmetry. Humans have evolved to see beauty in order and symmetry—think of the reliability of the rising and setting sun, the symmetrically mesmerizing forms of snowflakes or the famous studies that claim people find faces more attractive the more symmetrical they are.

While symmetry is generally taught early in school and most adults know what it is on sight, the mathematical definition is quite specific: symmetry is something or some process that remains the same even though something has changed. Thus, although the right and left sides of the face are two different sides, if the right side is (roughly—no face is wholly symmetrical) duplicated on the left side, humans will find that pleasing. It invites consistency and predictability.

This predictability is highly important to the scientific field. Science depends on relatively concise summaries of the world, and symmetry makes it easier to find those summaries. Symmetries also lend themselves to easier equations and models that can describe more things—this is why researchers have found the hunt for gravitational waves so seductive, because the theoretical equation for a gravitational wave's speed, wavelength and frequency would mirror that for light waves. Symmetry, in other words, simplifies the world and the universe down to a scale that is useful and practical for human comprehension and investigational tools.

Unsurprisingly, then, symmetry and asymmetry are key to discussing the scientific and mathematical concepts of order and chaos, because they represent these concepts in a visual form that is easy to process and understand. Chaos means the slow divergence of symmetry, eventually resulting in total asymmetry, while linear systems tend to be highly symmetrical. Fractals' relationship with symmetry is particularly interesting, because they have a special type of symmetry called scale invariance. This means that their symmetry exists in terms of scale, so that repetition of the pattern occurs at different scale levels (to test this, zoom in and out of a picture of a fractal). Scale invariance is also known as self-similarity, because the pattern is "similar" to itself mathematically at different scales. This produces a deep sense of visual order.

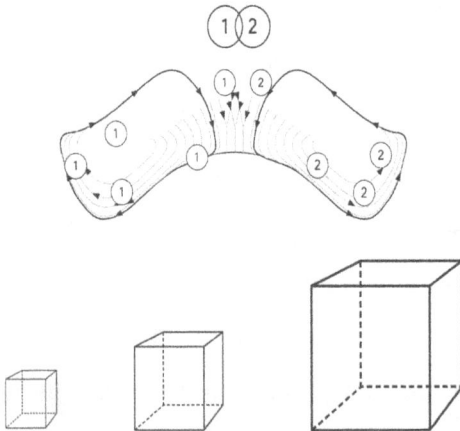

Picture 18: Scale Symmetry

The dimensions fractals occupy are called "fractal dimensions"—this is one of the few definitional points most mathematicians and scientists can agree on. The idea of recursive self-similarity developed by Mandelbrot is not a complete definition, and neither is the fractal dimension—fractal curves that fill space do not in fact have the dimensional requirements outlined by Mandelbrot. It is so difficult to find a satisfactory mathematical definition for fractals, in fact, that some mathematicians have argued there should be no exclusive, strict definition. Kenneth Falconer proposed that fractals should be a category that includes images and algorithms with some sort of self-similarity, detailed small-scale structures, patterns that are irregular in Euclidean (what we think of as "normal") geometry both locally and globally and some sort of recursive definition (that is, repeating back on itself). This broad definition helps exclude things that are self-similar but clearly not fractals, like straight lines, which are self-similar but meet none of the other requirements.[lvii]

Fractal-generating programs are the most common way to create fractals today, and as we previously learned are a popular subject in introductory-level computer science. There are several different types of these programs, including:

- Iterated function systems (IFS): These operate on the principle of repeated substitution, in which substitutions are iterated at a double rate for each iteration, building smaller and smaller scales of the pattern onto the initial starting pattern. These replacement iterations are geometric. This method can generate Sierpinski triangles and carpets, the Koch snowflake, the Cantor set and the Menger sponge, among other fractal curves.[lviii]

- Strange attractors: This method uses map or system solutions for chaotic initial-value differential or difference equations in its iterations. This can generate multifractal images.

- Lindenmayer systems, or L-systems: These generate fractals by running a branching function f, which is iterated recursively to create a branching fractal tree, rewriting over the previous string in the code with each iteration. This can help model veins in the human body, plants, cells and turtle graphics.

- Escape-time fractals: These either use a recurrence relation, which is a recursive equation, or a mathematical formula at each point in a space it occupies, enabling the construction of quasi-self-similar fractals over space such as the Mandelbrot set and the Julia set, which require two-dimensional vectors.

- Random fractals: These produce models of random motion or generation that follow fractal patterns such as Brownian motion, the Brownian tree and self-avoiding walks. These are used to model a variety of real life situations from natural landscapes to molecular and atomic motion.

- Finite subdivision rules: These use a topological algorithm that is recursive to model finite recursive processes such as cell division, generation of the Cantor set and the self-limiting Sierpinski carpet, which occurs within the bounds of an initial square.[lix]

Mathematical fractals

As you might have guessed, mathematical fractal generation depends on iterative processes. Iterations help perpetuate the feedback loops that generate fractals, but these iterations actually involve the repeated application of very simple geometry to a set of numbers and/or geometric forms. The famous "mental boss" fractal, for example, is generated through iterations baked into a simple function, which is applied to complex numbers.

Iterative equations often rely on a variable Y that is attached to a counter for each iteration, such as $n + 1$, which will thus generate a set of numbers from initial number n (which is generally zero). The function of Y will be something like $Y_n^3 + 5$, which will thus generate a new value for Y for each iteration. The Mandelbrot set is actually a representation of what we can think of as the properties of the set generated by an equation $Z_{n+1} = Z_n^2 + C$; it represents what happens at each starting value of C for the equation. The boundary of the Mandelbrot set image represents the value of C where Z shifts from staying finite to going to infinity. All this sounds very complex, but really just think of the image as describing what the equation does at different values.[lx]

An interesting visual property of fractals is that they have what we call "infinite variety," or infinite levels of detail and structure because of the scale invariance of fractals. Euclidean geometric shapes lose detail as one increases the scale ("zooms in")—think of zooming in very far on a square, to the point where it looks like a straight line. This also occurs with circles. However, fractals derive from iterative functions, which means exactly what it sounds like: the pattern continues to iterate as we "zoom in," therefore providing an infinite level of detail at every scale. You might have even encountered this in your daily life, if you find yourself standing between two mirrors; the reflections iterate infinitely in each other, creating a dizzying and fun effect! Because of this phenomenon, fractals can pack an infinite amount of form into a form (such as your repeating reflection being contained within the boundaries of the mirror), and they can also pack infinite length into a finite length.

The Koch curve, or Koch snowflake, is an excellent example of the iterative visual properties of fractals that we have previously examined in passing. To construct a Koch curve, start with an equilateral triangle. Then divide each line segment into three equal lengths, and draw another triangle for each line segment with the middle length as the base. Then, remove the middle length from each line segment. The first iteration produces the outline of a hexagram, but begins to resemble a very complex six-pointed snowflake through subsequent iterations, which can be infinite. In the case of infinite iterations, the length of the curve will be infinite despite looking quite defined in its path, due to its infinite variety. This is an example of infinite length being contained in finite length.[lxi]

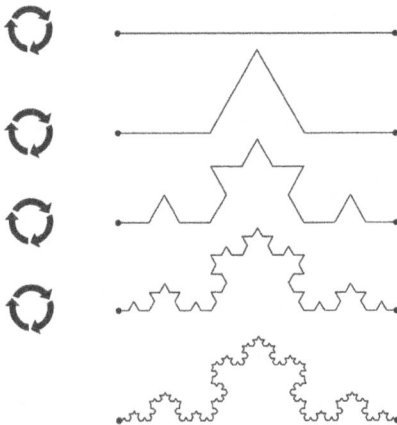

Picture 19: Koch Snowflake

Another example of a mathematical fractal is the Sierpinski fractal; the two famous subtypes of this fractal are the Sierpinski triangle and the Sierpinski carpet. Sierpinski fractals rely on the repeated iteration of pure geometry, which recurs into an existing shape. So, to create the Sierpinski triangle, we must first start with a simple black equilateral triangle, and then insert a while equilateral triangle whose points touch the midpoint of each side of the first triangle. One then simply repeats this process, causing a recursive geometry. Thus the number of triangles T can be expressed as $T = 3^n$, because each iteration results in three to the power of the number of iterations of black triangles. When $n = 1$, $T = 3$, and so on. The Sierpinski carpet is created using a similar method, where starting with a single black square, you divide it into 9 squares and remove the center. Repeat the process indefinitely with each of the 8 remaining black squares, and so on for as many iterations as one wants.[lxii]

Because of their scale invariance, fractals have a property called the scale-free property, wherein any scale can serve as a frame of reference and therefore there is no "correct" scale at which one could view the fractal. Linear systems tend to have a limited sense of scale; think back, again, to how a square looks like a line at too intimate of a scale. This sense of scale manifests itself as statistical "normality" of scale, wherein the features of the system, plotted on a graph according to the number of similar features, will show a normal distribution where the majority of features are similar to each other in a sort of qualitative mean, and few are outliers. Thus the features in the "normal" distribution are in the scale at which the linear system should be viewed.

The scale invariance we have mentioned in nonlinear fractal systems causes a lack of this normal distribution of features. Because there is no "proper" scale, there are "normal" features at all scale levels and therefore no specific, "normal" frame of reference for the fractal's viewing. Instead, the features' frequency of appearance is statistically distributed according to the power law. The power law is where there are a few big values, and therefore outliers, to one side of a distribution in a statistical graph, while there is a long tail of smaller values that stretches farther along the other side of the graph. This is because fractals tend to have only a few large features, but infinitely many smaller ones thanks to their recursive properties. Thus the power law graph for a fractal will have an infinite long tail. Statistical graphs are excellent visual representations of the differences between linear systems and fractals, because while linear systems have "normal" scale and normal scale distribution, fractals entirely lack both due to their lack of a "correct" scale.[lxiii]

Chapter 10: Real World Examples of Fractals

Now that we have a theoretical understanding of fractals, let's ground them in some more concrete examples. As previously mentioned, fractal images and sets are created through the repetition of extremely simple functions and rules. It is the repetition that provides fractals' complexity. Thus it is relatively easy to generate continuous branching fractals and spirals, because the repetition provides both order and complexity.

Branching fractals are exactly what they sound like. They build repetition on top of the existing forms in the fractal, creating a "fractal tree" that looks like the real thing. Leonardo da Vinci theorized that all branches of a tree are equivalent in thickness to the trunk, obeying the fractal property of scale invariance (though the concept was not invented yet at the time). This rule works better in theory than in practice (trees tend to be slightly mathematically imperfect), but it is a good way of understanding the principles behind fractal trees. Trees grow new limbs in a self-similar fashion, just as more theoretical fractals do. This growth is also recursive, as each new branch grows other branches. The points where branches grow are the bifurcations, to put this into terms previously discussed. Natural fractals such as trees, of course, are not infinite; that would not be possible in a universe where there is a finite amount of matter. However, what they lose in mathematical perfection they gain in comprehensibility as examples.[lxiv]

Rivers, as an example, are real life branching fractals that most people have encountered at least once in their lives. Earth's river ways form networks through fractal branching, which is facilitated by rainfall and erosion. The movement of water across the land carves out curved channels according to topography, which create a large space for water within a small space of land. This is, of course, highly similar to the fractals like the Koch snowflake that hold infinite length in the curve's length. River networks move constantly with a current precisely *because* they collect so much water; only a steady supply of water from the river's source can keep such a network going without constant rain. Interestingly, scientists have also found evidence of river networks on Mars, through the presence of dried fractal channels that mean Mars once had water, even rainfall.[lxv]

Lightning is another naturally occurring branching fractal. Mandelbrot wrote that lightning's path is inherently mathematically chaotic. This chaos creates fractal patterns in three dimensions as charged energy travels at the speed of light to Earth. Because lightning moves in a fractal pattern, the sound waves it generates—also known as thunder—also move in the same fractal pattern through space. These patterns can be captured in a lab on a small scale by supercharging and then discharging electrons made of an acrylic material. The electrons leave the pattern in which they discharged behind in the material that surrounds it, creating "frozen lightning" that has a clear fractal branching pattern. All lightning has this pattern, but neither human eyes nor photographs are able to capture the full fractal because it moves so quickly.[lxvi]

Spirals are another fractal that frequently appears in nature. Many plants, notably cacti, sunflowers, succulents, pinecones, Romanesco broccoli and daisies, contain leaves or petals that grow in a spiral fractal pattern. The nautilus builds its shell through literal three-dimensional fractal generation; the mollusk expands its shell by adding a new piece at a slightly larger scale and a slight rotation each time, generating a beautiful spiral fractal through repetition. Spirals with these same properties are found all over nature, and are also behind the "golden ratio" theory of art and aesthetics, which is still used by artists and filmmakers today.[lxvii]

Let's go back to Leonardo da Vinci's tree theory, which is now called Leonardo's Rule for Branches. Leonardo reached his idea with no mathematical concept of fractals; instead, he used logic to reach the conclusion based on the idea that branches acted as "plumbing" to move water and nutrients around the tree, and thus need to contain the same total area at all "scales" of the tree. You can replicate Leonardo's experiment yourself; if you go outside and observe some trees (try to make sure they're all the same species), divide each tree visually at each bifurcation point—where new branches occur. Do this to the fourth iteration of branching; we will call the trunk the "zero order," the first branching the "first order," and so on. Count the number of branches for each "order" and record it.

Leonardo da Vinci believed that one could predict the number of branches B at any given order n by using the simple equation $B = 2^n$, assuming each branch was the same size in a given order. Do your results follow this equation? Is the number of branches you've observed at each order consistent, which would imply a certain branching pattern for this species of tree? Try the same thing with trees of other species—do different species have different branching patterns? If so, why do you think this is true? All these questions will provide new ways to explore and think about branching fractals and their role in nature.[lxviii]

Fractals are also highly useful to the human body. Even if you are extremely tall, your body has to fit a *lot* of organs into it. The small intestine alone is twenty feet long, and the large intestine is an additional five feet. The circulatory system is even longer—end to end, all the veins, arteries and capillaries in your body would stretch for about 60,000 miles! Since the average human male is six feet tall, and the average human female five feet four inches, fractals are necessary just to be able to fit all the organs we need to stay alive into our own bodies.

Take, for example, the lungs. Casts of human lungs reveal an extremely intricate branching fractal pattern of blood vessels, all designed to oxygenate deoxygenated blood. Lungs are actually highly similar to trees in that the "branches" are necessary for the process of respiration, or moving oxygen and carbon dioxide through an organism. The relationship between branching fractals and respiration is an example of the Structure-Function Relationship. Scientists use this term to describe structures in organisms that look similar because they perform the same functions. Indeed, most organs exhibit this phenomenon in animals and plants. This is because the organs have the same structural needs; in this case, being able to have a *lot* of space to perform chemical reactions like respiration while being confined to a small bodily structure. Because respiration needs also exist based on scale (humans need to process more oxygen than mice), these fractals can exist at different scales for different organisms. Thus the amount of gas processed is directly proportional to surface area in the fractal. Again, because these fractals are natural, they are finite, but they still contain a stunningly large surface area—in humans, roughly the area of a tennis court.[lxix]

If you are interested in the mathematics behind this phenomenon, you can calculate something called the surface area to volume ratio. For a simple geometric solid like a sphere, this is easy to calculate; just divide the surface area formula, $SA = 4\pi r^2$, by the volume formula, $V = 4/3\pi r^3$, to arrive at the ratio of $3/r$ where r is the radius of the sphere. This is an inverse relationship; as the sphere gets larger, with a larger radius, the surface area to volume ratio becomes smaller. If you plot this equation on a graph, you will see that it is a hyperbola.

To calculate the surface area to volume ratio of the lungs, assume the lungs' surface area is $100 \ m^2$ (the tennis court area mentioned above) and the volume is $.005 \ m^3$ (or 5 liters). As you will see, this ratio is quite large! Now, set the formula for the ratio for a sphere as equal to the number you got. Then calculate for the size of the radius. As you can see, the sphere would have to be quite small to have the same surface area to volume ratio! This vividly illustrates just how useful fractals are; if our lungs did not have a fractal structure, they would need to be impossibly large.

Blood vessels are also branching fractals. Fractals come in useful here because we not only have an extremely large amount of blood to circulate around our bodies, we also have to have a blood vessel near every single cell in our bodies, or else they will die, taking us with them. Thus we have roughly 150,000 kilometers of blood vessels in our bodies! Most of these kilometers, however, are covered by capillaries that are so tiny they have to be seen by a microscope—the logical result of many scales of branching fractals in the circulatory system. Thus blood vessels can cover the massive area and transport the massive amount of material they need to while being confined to our relatively small bodies.[lxx]

Our nervous system also exhibits fractal branching, for similar reasons to the circulatory system; our neurons need to transport a lot of information to a *lot* of different nerve endings. Anyone who has ever suffered a pinched nerve or sciatica will understand how crucial these branching structures are to making it through our day! The locus for these fractals is, of course, the brain, which has billions of neurons that generate trillions of synapses. These neurons are all branching fractals as well; this allows for maximum space efficiency within the brain and maximum contact with other neurons, which is crucial for correct synapse firing. Without all our synapses firing, humans suffer severe neurological problems, even brain death or total death. Our neurons need to be able to communicate with each other! Because the human body's organs—especially the brain—are so complex yet small, branching fractals are absolutely necessary to our survival.[lxxi]

Finally, enjoy these beautiful images of fractals:

Picture 20: A Fern Branching[lxxii]

Picture 21: A Tree Branching[lxxiii]

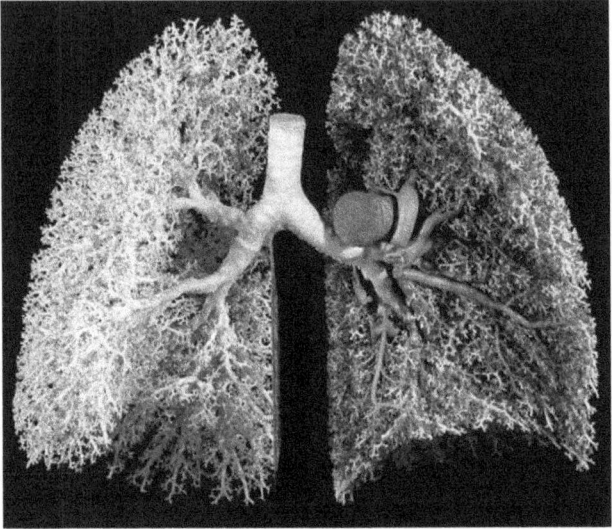

Picture 22: A Lung Branching[lxxiv]

Picture 23: Spiral Fractal[lxxv]

Conclusion

For centuries, scientists and mathematicians relied on the classical modes of their disciplines. These were the physics of Newton and the geometry of Euclid. They had clear laws and clear rules that worked most of the time. They described a universe that was neat and orderly, where everything was organized and perfected by some larger purpose. But even then, there were anomalies. There were phenomena that didn't fit in this ordered universe, that didn't quite generate the expected data. Some brushed them off as experimental flaws, but they surely haunted others—why? Shouldn't the whole universe follow the laws of physics? Is that not why they are called laws?

The organizing principle behind classical mathematical and scientific disciplines was the Platonic ideal of order and harmony. But there is a reason we now call this an "ideal": the world is rarely an ordered, harmonic place. There are capricious changes in weather. Hurricanes can change course and skip a town it was supposed to level while erasing another off the map. Few see a market crash coming until it's too late. A healthy woman can die of a sudden aneurysm. Celestial bodies appear to deviate from their set orbits. The universe has always been a chaotic place. There has always been anomaly in nature; it is how it is built.

Chaos theory, then, was the next great step in scientific study. It provided answers to the stubborn *Why?* that had dogged so many scientists, economists, mathematicians, engineers, doctors, and others for generations. It evolved over many generations, of course.

The accumulation of anomalies and theories over the centuries led to an explosion of scientific advancement in the 1960s and 70s that was then accelerated by the rapid development of the computer. Edward Lorenz and Benoit Mandelbrot advanced chaos theory by leaps and bounds through the simple observation that small differences in input data could result in wildly different results in weather and economic models. From there, these ideas spread to doctors who suddenly found sense in the human body's more chaotic behaviors, to biologists who began to realize that small changes in the ecosystem could wipe out entire species, to climate scientists who realized that small changes in carbon levels— let alone the massive ones that occurred in the late twentieth century—could permanently alter the Earth's climate and habitability. Once humans recognized the root of chaos, suddenly all the chaos in nature began to make sense. The path of lightning, the structure of the nervous system, even the fact that one painting seemed more beautiful than another all began to make more sense. At the root of all of them was chaos.

The beauty of chaos was not only its strange order: it is also how it has emerged as an organizing principle uniting so many disparate fields. The connections between mathematics and physics or economics are easy to understand, but the idea that human bodies and human behavior follow the same principles is a new and strangely beautiful idea. Chaos theory has helped facilitate disciplinary cross-pollination in a way few other scientific theories have been able to. Even popular children's movies are now influenced by chaotic systems algorithms now.

Still, chaos theory remains a bit of a scientific mystery. It has only entered the popular lexicon in the last twenty years or so, and many still have only a vague idea of what it means. It has developed by leaps and bounds with the development of advanced computing techniques and algorithms, which are now able to do significantly more complex calculations thousands of times faster than they were in the 1960s, when Lorenz noticed a mistake in his data printout. It is an exotic theory, and still being mathematically fleshed out—indeed, this is why topology is such a trending topic among young mathematicians engaging in doctoral studies today. Much remains to be learned, a prospect that many scientists find exciting but a bit frightening. For example, what developments could come when quantum computing becomes mainstream? How will it develop within quantum physics, which is also still a relatively young field? Will the continued development of chaos theory completely destabilize everything we thought we knew about the universe?

Astrophysicists have often claimed that the more we learn about the universe in which we live, the more mysterious it becomes. Chaos theory fits with this mysteriousness. It initially seems almost threatening, the idea that small actions can have unforeseen large effects. The butterfly effect is often understood in popular culture as some sort of menacing force, threatening to topple our lives with the most insignificant, thoughtless decision. Each choice, whether to eat lunch at a certain cafe or to walk through the park, can turn dizzying when it becomes a bifurcation point of reality. People have not changed much since Plato's time; we still don't love living in a world we can never control.

The idea of the world as a linear system is certainly appealing. Linear systems are easy to predict. They conform to highly simple mathematical models. They can be added together; their output is always in direct proportion to the input. They help us explain the best-case scenario, the most mathematically perfect outcome. But linear systems rarely occur in real life. They are, quite simply, too good to be true. The physical world is too unpredictable to be one hundred percent explained by idealized formulae.

Nonlinear systems are not as easy to model. They're not mathematically simple or organized. They require complex algorithms and exceptions to rules, and their results will always interact with other nonlinear systems' results. They are messy. They have an impact on each other. Their defining characteristics are synergy, which requires both differentiation of "labor" and coordination, and interference, which means that results are often diminished due to interaction. All this is to say that nonlinear systems do not behave as an island, and thus are difficult mathematical beasts. However, chaos theory lends order to this disorder. It helps to explain why nonlinear systems' interactions produce unpredictable results.

The problem of unpredictability is only compounded in dynamic systems; that is why, instead of sure outcomes for dynamic systems such as the weather or the economy, we can only give ballpark predictions as opposed to absolute certainties about behavior. These systems change over time, and with these changes come new attractors, interactions, and possible feedback loops that can all affect the outcome. This is not to say that these outcomes are random; in fact, quite the opposite. Nonlinear dynamic systems still behave the way they do as a result of the initial conditions and the subsequent interactions within the system, which rarely if never come out of the blue. That is why it *is* possible to make estimates about dynamic systems' behavior. It is just important to take their full complexity into account.

Chaos is often conflated with random behavior in everyday usage. However, chaos theory does not deal with random behavior at all. Instead, it helps us understand systems that are a little less predictable, that do not have outputs in direct proportion to their inputs, that seem to have results that make no sense. Chaos is not a response that makes no sense; it is instead a systemic response that makes little linear sense for the input. Chaos produces beautiful things, like fractals, which build upon themselves to create infinite space, infinite line, within that which is finite. They have their own kind of chaotic symmetry that does not have to be linear. Chaos theory does not have to be complex, even though it helps to explain complex mathematical phenomena. Like fractals, it can be astoundingly simple in how it articulates the quirks of nonlinear dynamic systems.

Life, after all, is a system. It is elements which are interconnected to serve a purpose or function. Many people think that "purpose" means that they must make some grand accomplishment or change the world—end hunger, bring world peace, et cetera. But this is not what a system is at all. As we've learned, a mathematical system is simply something that happens. It is when multiple parts interact and produce a result. The result doesn't have to be "right." It doesn't even have to be predictable. This is the beauty of chaos. It doesn't require perfect order. It doesn't require everything to go according to plan. Instead, it allows for the system to function, with strange attractors, runaway feedback loops, phase transitions and all. Perhaps the true secret to the "meaning of life" is the exact same answer that physicists and mathematicians and meteorologists have found: chaos. Perhaps our meaning is just to interact in our own small ways and find our own strange attractors. Perhaps that is how we have a big effect on the world, what people would call a legacy—embracing the chaos.

A.R.

Reference

Abramovitz, Mortimer. Davidson, Michael W. Interference. Olympus. 2019. https://www.olympus-lifescience.com/en/microscope-resource/primer/lightandcolor/interference/

Addison, Paul S. (1997). Fractals and Chaos: An Illustrated Course. Institute of Physics. p. 19. ISBN 0-7503-0400-6.

Blanchard, P.; Devaney, R. L.; Hall, G. R. Differential Equations. London: Thompson. pp. 96–111. ISBN 978-0-495-01265-8. 2006.

Boeing, G. Visual Analysis of Nonlinear Dynamic Systems: Chaos, Fractals, Self-Similarity and the Limits of Prediction.

Systems. 4 (4): 37. arXiv:1608.04416. doi:10.3390/systems4040037. 2016.

Boeing (2015). "Chaos Theory and the Logistic Map". Journal of the Optical Society of America B Optical Physics. 3 (5): 741. Retrieved 2015-07-16.

Boeing, Geoff. Visual Analysis of Nonlinear Dynamic Systems: Chaos, Fractals, Self-Similarity and the Limits of Prediction. Department of City and Regional Planning, University of California. 2016. https://arxiv.org/pdf/1608.04416.pdf

Bonhoeffer, Sebastian. The logistic difference equation and the route to chaotic behavior. Institute of Integrative Biology ETH Zurich. 2019. https://ethz.ch/content/dam/ethz/special-interest/usys/ibz/theoreticalbiology/education/learningmaterials/701-1424-00L/lde.pdf

Briggs, John (1992). Fractals:The Patterns of Chaos. London: Thames and Hudson. p. 148. ISBN 978-0-500-27693-8.

Brunori, Paola; Magrone, Paola; Lalli, Laura Tedeschini (2018-07-07), "Imperial Porphiry and Golden Leaf: Sierpinski Triangle in a Medieval Roman Cloister", Advances in Intelligent Systems and Computing, Springer International Publishing, pp. 595–609, doi:10.1007/978-3-319-95588-9_49, ISBN 9783319955872

Carl Shapiro and Hal R. Varian. Information Rules. Harvard Business Press. ISBN 978-0-87584-863-1. 1999.

Cengage Learning. Modeling With Exponential And Logarithmic Functions. Cengage Learning. 2005. http://www.math.yorku.ca/~raguimov/math151

0_y13/PreCalc6_04_06%20%5BCompatibility
%20Mode%5D.pdf

De Canete, Javier, Cipriano Galindo, and
Inmaculada Garcia-Moral. System Engineering
and Automation: An Interactive Educational
Approach. Berlin: Springer. p. 46. ISBN 978-
3642202292. 2011.

Diacu, Florin; Holmes, Philip (1996). Celestial
Encounters: The Origins of Chaos and
Stability. Princeton University Press.

Dictionary. Synergy. Dictionary. 2019.
https://www.dictionary.com/browse/synergy
Dizikes, Petyer (2008). "The meaning of the
butterfly". The Boston Globe. Archived from
the original on 18 April 2016. Retrieved 8 June
2016.

Eisler, Z.; Bartos, I.; Kertész, J. (2008).
"Fluctuation scaling in complex systems:

Taylor's law and beyond". Adv Phys. 57 (1): 89–142. arXiv:0708.2053. Bibcode:2008AdPhy..57...89E. doi:10.1080/00018730801893043.

Example extracted from: Algebra 1. Solving Real-World Problems Using Linear Systems. Algebra 1. 2019. https://students.ga.desire2learn.com/d2l/lor/viewer/viewFile.d2lfile/1798/12632/Algebra_ReasoningWithEquationsandInequalities12.html

Encyclopedia. Normal Distribution. Encyclopedia. 2019. https://www.encyclopedia.com/science-and-technology/mathematics/mathematics/normal-distribution#3

F.H. Busse, Transition to turbulence in Rayleigh - Bénard convection, In Hydrodynamic Instabilities and the Transition to Turbulence, eds. H.L. Swinney and J.P.

Gollub (Topics in Appl. Phys., vol. 45). 2nd edition. Berlin, Springer. 1985. pp. 97 - 137.

Fractal Foundation. Branching Fractals. Fractal Foundation. 2019.
https://fractalfoundation.org/OFC/OFC-1-1.html

Fractal Foundation. Fractal Blood Vessels. Fractal Foundation. 2019.
https://fractalfoundation.org/OFC/OFC-1-3.html

Fractal Foundation. Fractals in the body. Fractal Foundation. 2019.
https://fractalfoundation.org/OFC/OFC-1-2.html

Fractal Foundation. Fractal Lightning. Fractal Foundation. 2019.

https://fractalfoundation.org/OFC/OFC-1-5.html

Fractal Foundation. Fractal Neurons. Fractal Foundation. 2019. https://fractalfoundation.org/OFC/OFC-1-6.html

Fractal Foundation. Fractal Rivers. Fractal Foundation. 2019. https://fractalfoundation.org/OFC/OFC-1-4.html

Fractal Foundation. From the Simple to the Complex. Fractal Foundation. 2019. https://fractalfoundation.org/OFC/OFC-4-1.html

Fractal Foundation. Spirals. Fractal Foundation. 2019. https://fractalfoundation.org/OFC/OFC-1-7.html

Frame, Angus (August 3, 1998). "Iterated Function Systems". In Pickover, Clifford A. (ed.). Chaos and fractals: a computer graphical journey : ten year compilation of advanced research. Elsevier. pp. 349–351. ISBN 978-0-444-50002-1.

Gintautas, V. Resonant forcing of nonlinear systems of differential equations. Chaos. 18 (3): 033118. arXiv:0803.2252. Bibcode:2008. Chaos.18c3118G. doi:10.1063/1.2964200. PMID 19045456. 2008.

Gleick, James (1987). Chaos: Making a New Science. London: Cardinal. p. 17. ISBN 978-0-434-29554-8.

Gutzwiller, Martin C. (1990). Chaos in Classical and Quantum Mechanics. New York: Springer-Verlag. ISBN 0-387-97173-4.

Heeger, David. Linear System Theory. Department of Psychology, New York University. 2007. http://www.cns.nyu.edu/~david/handouts/linear -systems/linear-systems.html

Jaeger, Gregg. "The Ehrenfest Classification of Phase Transitions: Introduction and Evolution". Archive for History of Exact Sciences. 53 (1): 51–81. 1998. doi:10.1007/s004070050021

Karperien, Audrey (2004). Defining microglial morphology: Form, Function, and Fractal Dimension. Charles Sturt University. doi:10.13140/2.1.2815.9048.

Kellert, Stephen H. (1993). In the Wake of Chaos: Unpredictable Order in Dynamic Systems. University of Chicago Press. p. 32. ISBN 978-0-226-42976-2.

Koba, Mark. Market Circuit Breakers: CNBC Explains. CNBC. 2011. https://www.cnbc.com/id/44059883

Lorenz, Edward N. "Deterministic Nonperiodic Flow". Journal of the Atmospheric Sciences. 20 (2): 130–141. Bibcode:1963JAtS...20..130L. doi:10.1175/1520-0469(1963)020<0130:dnf>2.0.co. 1963.

Lynsisneuro. Linear Systems and the Superposition Principle. Linsysneuro. 2013. https://linsysneuro.wordpress.com/2013/02/22/linear-systems-and-the-superposition-principle/

Manage. Punctuated Equilibrium Model. Manage. 2019. https://www.kbmanage.com/concept/punctuated-equilibrium-model

Markov, A. A. (1954). Theory of Algorithms. [Translated by Jacques J. Schorr-Kon and PST staff] Imprint Moscow, Academy of Sciences of the USSR, 1954 [Jerusalem, Israel Program for Scientific Translations, 1961; available from Office of Technical Services, United States Department of Commerce] Added t.p. in Russian Translation of Works of the Mathematical Institute, Academy of Sciences of the USSR, v. 42. Original title: Teoriya algorifmov. [QA248.M2943 Dartmouth College library. U.S. Dept. of Commerce, Office of Technical Services, number OTS 60-51085.]

Math Planet. Graphing Linear Systems. Math Planet. 2019. https://www.mathplanet.com/education/algebra-1/systems-of-linear-equations-and-inequalities/graphing-linear-systems

Mandelbrot, Benoît (1963). "The variation of certain speculative prices". Journal of Business. 36 (4): 394–419. doi:10.1086/294632. JSTOR 2350970.

Mandelbrot, Benoît (5 May 1967). "How Long Is the Coast of Britain? Statistical Self-Similarity and Fractional Dimension". Science. 156 (3775): 636–8. Bibcode:1967Sci...156..636M. doi:10.1126/science.156.3775.636. PMID 17837158

Meadows, Donella. Thinking in Systems. Chelsea Green Publishing. 2008.

M.E.J. Newman. Power laws, Pareto distributions and Zipf's law. Contemporary Physics, Vol. 46, No. 5, September–October 2005, 323 – 351. 2004. http://tuvalu.santafe.edu/~aaronc/courses/7000/readings/Newman_05_PowerLawsParetoDistributionsAndZipfsLaw.pdf

Motter, A. E.; Campbell, D. K. (2013). "Chaos at fifty". Phys. Today. 66 (5): 27–33.

arXiv:1306.5777.

Bibcode:2013PhT....66e..27M.

doi:10.1063/pt.3.1977.

On Digital Marketing. Social Network Theory
and Metcalfe's Law. On Digital Marketing.
2018.
https://ondigitalmarketing.com/learn/odm/foun
dations/social-network-theory-and-metcalfes-
law/

Picture 20 was retrieved from
https://www.etsy.com/hk-
en/listing/203934132/vintage-botanical-fern-
temporary-tattoo

Picture 21 was retrieved from
https://packetpushers.net/make-like-tree-
branch/

Picture 22 was retrieved from
https://fractalfoundation.org/OFC/OFC-1-
2.html

Picture 23 was retrieved from
https://pixabay.com/illustrations/fractal-spiral-
endless-mathematics-199054/

Poincaré, J. Henri (2017). The three-body
problem and the equations of dynamics :
Poincaré's foundational work on dynamic
systems theory. Popp, Bruce D. (Translator).
Cham, Switzerland: Springer International
Publishing. ISBN 9783319528984. OCLC
987302273.

Pooja Patnaik. Jeremy Jostad. Dynamic
Systems Theory. Psychology University of
Utah. 2019.
http://old.psych.utah.edu/~jb4731/systems/Dyn
amicSystemsIntro.html

Pooja Patnaik. Jeremy Jostad. Dynamic Systems Theory. Psychology University of Utah. 2019.
http://old.psych.utah.edu/~jb4731/systems/DynamicSystemsIntro.html

Ross D. Arnold, Jon P. Wade. A Definition of Systems Thinking: A Systems Approach. Elsevier. 2015.
http://creativecommons.org/licenses/by-nc-nd/4.0/

Spacey, John. What is a long tail disctribution? Simplicable. 2017.
https://simplicable.com/new/long-tail-distribution

Sprott, Julien Clinton (2003). Chaos and Time-Series Analysis. Oxford University Press. ISBN 978-0-19-850840-3.

Systems Academy. Homogeneity Principle. Systems Academy. 2019. https://systemsacademy.io/glossary/homogeneity-principle/

The Penguin Dictionary of Physics, ed. Valerie Illingworth. Penguin Books, London. 1991.

Trochet, Holly (2009). "A History of Fractal Geometry". MacTutor History of Mathematics. Archived from the original on February 4, 2012.

Vicsek, Tamás (1992). Fractal growth phenomena. Singapore/New Jersey: World Scientific. pp. 31, 139–146. ISBN 978-981-02-0668-0.

Wikipedia. Exponentiation. Wikipedia. 2019. https://en.wikipedia.org/wiki/Exponentiation#cite_note-MacTutor-1

Wikipedia. Iteration. Wikipedia. 20 19.
https://en.wikipedia.org/wiki/Iteration

Wikipedia. Phase Transition. Wikipedia. 2019.
https://en.wikipedia.org/wiki/Phase_transition

Xiannong, Meng. Discrete and Continuous
Systems. Bucknell. 2002.
https://www.eg.bucknell.edu/~xmeng/Course/C
S6337/Note/master/node6.html

Yong Wang. Kwok-Wo Wong. Xiaofeng Liao.
Tao Xian. Guanrong, Chen. A chaos-based
image encryption algorithm with variable
control parameters. Chaos, Solitons & Fractals,
Volume 41, Issue 4, Pages 1773-178. 2009.
https://doi.org/10.1016/j.chaos.2008.07.031

Endnotes

[i] Meadows, Donella. Thinking in Systems. Chelsea Green Publishing. 2008.

[ii] Ross D. Arnold, Jon P. Wade. A Definition of Systems Thinking: A Systems Approach. Elsevier. 2015.
http://creativecommons.org/licenses/by-nc-nd/4.0/

[iii] Ross D. Arnold, Jon P. Wade. A Definition of Systems Thinking: A Systems Approach. Elsevier. 2015.
http://creativecommons.org/licenses/by-nc-nd/4.0/

[iv] The Penguin Dictionary of Physics, ed. Valerie Illingworth. Penguin Books, London. 1991.

[v] Heeger, David. Linear System Theory. Department of Psychology, New York University. 2007.
http://www.cns.nyu.edu/~david/handouts/linear-systems/linear-systems.html

[vi] Systems Academy. Homogeneity Principle. Systems Academy. 2019.
https://systemsacademy.io/glossary/homogeneity-principle/

[vii] Math Planet. Graphing Linear Systems. Math Planet. 2019. https://www.mathplanet.com/education/algebra-1/systems-of-linear-equations-and-inequalities/graphing-linear-systems

[viii] Lynsisneuro. Linear Systems and the Superposition Principle. Linsysneuro. 2013. https://linsysneuro.wordpress.com/2013/02/22/linear-systems-and-the-superposition-principle/

[ix] Example extracted from: Algebra 1. Solving Real-World Problems Using Linear Systems. Algebra 1. 2019. https://students.ga.desire2learn.com/d2l/lor/viewer/viewFile.d2lfile/1798/12632/Algebra_ReasoningWithEquationsandInequalities12.html

[x] Example extracted from: Algebra 1. Solving Real-World Problems Using Linear Systems. Algebra 1. 2019. https://students.ga.desire2learn.com/d2l/lor/viewer/viewFile.d2lfile/1798/12632/Algebra_ReasoningWithEquationsandInequalities12.html

[xi] Boeing, G. Visual Analysis of Nonlinear Dynamic Systems: Chaos, Fractals, Self-Similarity and the Limits of Prediction. Systems. 4 (4): 37. arXiv:1608.04416. doi:10.3390/systems4040037. 2016.

[xii] De Canete, Javier, Cipriano Galindo, and Inmaculada Garcia-Moral. System

Engineering and Automation: An Interactive Educational Approach. Berlin: Springer. p. 46. ISBN 978-3642202292. 2011.

[xiii] Gintautas, V. Resonant forcing of nonlinear systems of differential equations. Chaos. 18 (3): 033118. arXiv:0803.2252. Bibcode:2008. Chaos.18c3118G. doi:10.1063/1.2964200. PMID 19045456. 2008.

[xiv] Dictionary. Synergy. Dictionary. 2019. https://www.dictionary.com/browse/synergy

[xv] Abramovitz, Mortimer. Davidson, Michael W. Interference. Olympus. 2019. https://www.olympus-lifescience.com/en/microscope-resource/primer/lightandcolor/interference/

[xvi] Markov, A. A. (1954). Theory of Algorithms. [Translated by Jacques J. Schorr-Kon and PST staff] Imprint Moscow, Academy of Sciences of the USSR, 1954 [Jerusalem, Israel Program for Scientific Translations, 1961; available from Office of Technical Services, United States Department of Commerce] Added t.p. in Russian Translation of Works of the Mathematical Institute, Academy of Sciences of the USSR, v. 42. Original title: Teoriya algorifmov. [QA248.M2943 Dartmouth College library. U.S. Dept. of Commerce,

Office of Technical Services, number OTS 60-51085.]

[xvii] Wikipedia. Exponentiation. Wikipedia. 2019.
https://en.wikipedia.org/wiki/Exponentiation#cite_note-MacTutor-1

[xviii] Wikipedia. Iteration. Wikipedia. 20 19.
https://en.wikipedia.org/wiki/Iteration

[xix] Cengage Learning. Modeling With Exponential And Logarithmic Functions. Cengage Learning. 2005.
http://www.math.yorku.ca/~raguimov/math1510_y13/PreCalc6_04_06%20%5BCompatibility%20Mode%5D.pdf

[xx] M.E.J. Newman. Power laws, Pareto distributions and Zipf's law. Contemporary Physics, Vol. 46, No. 5, September–October 2005, 323 – 351. 2004.
http://tuvalu.santafe.edu/~aaronc/courses/7000/readings/Newman_05_PowerLawsParetoDistributionsAndZipfsLaw.pdf

[xxi] On Digital Marketing. Social Network Theory and Metcalfe's Law. On Digital Marketing. 2018.
https://ondigitalmarketing.com/learn/odm/foundations/social-network-theory-and-metcalfes-law/

[xxii] Carl Shapiro and Hal R. Varian. Information Rules. Harvard Business Press. ISBN 978-0-87584-863-1. 1999.

xxiii Encyclopedia. Normal Distribution. Encyclopedia. 2019. https://www.encyclopedia.com/science-and-technology/mathematics/mathematics/normal-distribution#3

xxiv Koba, Mark. Market Circuit Breakers: CNBC Explains. CNBC. 2011. https://www.cnbc.com/id/44059883

xxv Spacey, John. What is a long tail disctribution? Simplicable. 2017. https://simplicable.com/new/long-tail-distribution

xxvi Pooja Patnaik. Jeremy Jostad. Dynamic Systems Theory. Psychology University of Utah. 2019. http://old.psych.utah.edu/~jb4731/systems/DynamicSystemsIntro.html

xxvii F.H. Busse, Transition to turbulence in Rayleigh - Bénard convection, In Hydrodynamic Instabilities and the Transition to Turbulence, eds. H.L. Swinney and J.P. Gollub (Topics in Appl. Phys., vol. 45). 2nd edition. Berlin, Springer. 1985. pp. 97 - 137.

xxviii Bonhoeffer, Sebastian. The logistic difference equation and the route to chaotic behavior. Institute of Integrative Biology ETH Zurich. 2019. https://ethz.ch/content/dam/ethz/special-interest/usys/ibz/theoreticalbiology/educat

ion/learningmaterials/701-1424-
00L/lde.pdf
xxix Boeing, Geoff. Visual Analysis of
Nonlinear Dynamic Systems: Chaos, Fractals,
Self-Similarity and the Limits of Prediction.
Department of City and Regional Planning,
University of California. 2016.
https://arxiv.org/pdf/1608.04416.pdf
xxx Wikipedia. Phase Transition. Wikipedia.
2019.
https://en.wikipedia.org/wiki/Phase_transit
ion
xxxi Yong Wang. Kwok-Wo Wong. Xiaofeng
Liao. Tao Xian. Guanrong, Chen. A chaos-
based image encryption algorithm with
variable control parameters. Chaos, Solitons
& Fractals, Volume 41, Issue 4, Pages 1773-
178. 2009.
https://doi.org/10.1016/j.chaos.2008.07.03
1
xxxii Pooja Patnaik. Jeremy Jostad. Dynamic
Systems Theory. Psychology University of
Utah. 2019.
http://old.psych.utah.edu/~jb4731/systems
/DynamicSystemsIntro.html
xxxiii Xiannong, Meng. Discrete and
Continuous Systems. Bucknell. 2002.
https://www.eg.bucknell.edu/~xmeng/Cou
rse/CS6337/Note/master/node6.html
xxxiv Boeing, G. Visual Analysis of Nonlinear
Dynamic Systems: Chaos, Fractals, Self-

Similarity and the Limits of Prediction. Systems. 4 (4): 37. arXiv:1608.04416. doi:10.3390/systems4040037. 2016.

xxxv Boeing, G. Visual Analysis of Nonlinear Dynamic Systems: Chaos, Fractals, Self-Similarity and the Limits of Prediction. Systems. 4 (4): 37. arXiv:1608.04416. doi:10.3390/systems4040037. 2016.

xxxvi Pooja Patnaik. Jeremy Jostad. Dynamic Systems Theory. Psychology University of Utah. 2019. http://old.psych.utah.edu/~jb4731/systems/DynamicSystemsIntro.html

xxxvii Pooja Patnaik. Jeremy Jostad. Dynamic Systems Theory. Psychology University of Utah. 2019. http://old.psych.utah.edu/~jb4731/systems/DynamicSystemsIntro.html

xxxviii Boeing (2015). "Chaos Theory and the Logistic Map". Journal of the Optical Society of America B Optical Physics. 3 (5): 741. Retrieved 2015-07-16.

xxxix Kellert, Stephen H. (1993). In the Wake of Chaos: Unpredictable Order in Dynamic Systems. University of Chicago Press. p. 32. ISBN 978-0-226-42976-2.

xl Poincaré, J. Henri (2017). The three-body problem and the equations of dynamics : Poincaré's foundational work on dynamic systems theory. Popp, Bruce D. (Translator). Cham, Switzerland: Springer International

Publishing. ISBN 9783319528984. OCLC 987302273.

[xli] Diacu, Florin; Holmes, Philip (1996). Celestial Encounters: The Origins of Chaos and Stability. Princeton University Press.

[xlii] Sprott, Julien Clinton (2003). Chaos and Time-Series Analysis. Oxford University Press. ISBN 978-0-19-850840-3.

[xliii] Gleick, James (1987). Chaos: Making a New Science. London: Cardinal. p. 17. ISBN 978-0-434-29554-8.

[xliv] Mandelbrot, Benoît (1963). "The variation of certain speculative prices". Journal of Business. 36 (4): 394–419. doi:10.1086/294632. JSTOR 2350970.

[xlv] Mandelbrot, Benoît (5 May 1967). "How Long Is the Coast of Britain? Statistical Self-Similarity and Fractional Dimension". Science. 156 (3775): 636–8. Bibcode:1967Sci...156..636M. doi:10.1126/science.156.3775.636. PMID 17837158

[xlvi] Motter, A. E.; Campbell, D. K. (2013). "Chaos at fifty". Phys. Today. 66 (5): 27–33. arXiv:1306.5777. Bibcode:2013PhT....66e..27M. doi:10.1063/pt.3.1977.

[xlvii] Boeing, Geoff. Visual Analysis of Nonlinear Dynamic Systems: Chaos, Fractals, Self-Similarity and the Limits of Prediction. Department of City and Regional Planning,

University of California. 2016.
https://arxiv.org/pdf/1608.04416.pdf
[xlviii] Lorenz, Edward N. "Deterministic Nonperiodic Flow". Journal of the Atmospheric Sciences. 20 (2): 130–141. Bibcode:1963JAtS...20..130L. doi:10.1175/1520-0469(1963)020<0130:dnf>2.0.co. 1963.
[xlix] Gutzwiller, Martin C. (1990). Chaos in Classical and Quantum Mechanics. New York: Springer-Verlag. ISBN 0-387-97173-4.
[l] Dizikes, Petyer (2008). "The meaning of the butterfly". The Boston Globe. Archived from the original on 18 April 2016. Retrieved 8 June 2016.
[li] Jaeger, Gregg. "The Ehrenfest Classification of Phase Transitions: Introduction and Evolution". Archive for History of Exact Sciences. 53 (1): 51–81. 1998. doi:10.1007/s004070050021
[lii] Blanchard, P.; Devaney, R. L.; Hall, G. R. Differential Equations. London: Thompson. pp. 96–111. ISBN 978-0-495-01265-8. 2006.
[liii] Manage. Punctuated Equilibrium Model. Manage. 2019.
https://www.kbmanage.com/concept/punctuated-equilibrium-model
[liv] Briggs, John (1992). Fractals:The Patterns of Chaos. London: Thames and Hudson. p. 148. ISBN 978-0-500-27693-8.

[lv] Trochet, Holly (2009). "A History of Fractal Geometry". MacTutor History of Mathematics. Archived from the original on February 4, 2012.

[lvi] Trochet, Holly (2009). "A History of Fractal Geometry". MacTutor History of Mathematics. Archived from the original on February 4, 2012.

[lvii] Karperien, Audrey (2004). Defining microglial morphology: Form, Function, and Fractal Dimension. Charles Sturt University. doi:10.13140/2.1.2815.9048.

[lviii] Frame, Angus (August 3, 1998). "Iterated Function Systems". In Pickover, Clifford A. (ed.). Chaos and fractals: a computer graphical journey : ten year compilation of advanced research. Elsevier. pp. 349–351. ISBN 978-0-444-50002-1.

[lix] Vicsek, Tamás (1992). Fractal growth phenomena. Singapore/New Jersey: World Scientific. pp. 31, 139–146. ISBN 978-981-02-0668-0.

[lx] Fractal Foundation. From the Simple to the Complex. Fractal Foundation. 2019. https://fractalfoundation.org/OFC/OFC-4-1.html

[lxi] Addison, Paul S. (1997). Fractals and Chaos: An Illustrated Course. Institute of Physics. p. 19. ISBN 0-7503-0400-6.

[lxii] Brunori, Paola; Magrone, Paola; Lalli, Laura Tedeschini (2018-07-07), "Imperial

Porphiry and Golden Leaf: Sierpinski Triangle in a Medieval Roman Cloister", Advances in Intelligent Systems and Computing, Springer International Publishing, pp. 595–609, doi:10.1007/978-3-319-95588-9_49, ISBN 9783319955872

[lxiii] Eisler, Z.; Bartos, I.; Kertész, J. (2008). "Fluctuation scaling in complex systems: Taylor's law and beyond". Adv Phys. 57 (1): 89–142. arXiv:0708.2053. Bibcode:2008AdPhy..57...89E. doi:10.1080/00018730801893043.

[lxiv] Fractal Foundation. Branching Fractals. Fractal Foundation. 2019. https://fractalfoundation.org/OFC/OFC-1-1.html

[lxv] Fractal Foundation. Fractal Rivers. Fractal Foundation. 2019. https://fractalfoundation.org/OFC/OFC-1-4.html

[lxvi] Fractal Foundation. Fractal Lightning. Fractal Foundation. 2019. https://fractalfoundation.org/OFC/OFC-1-5.html

[lxvii] Fractal Foundation. Spirals. Fractal Foundation. 2019. https://fractalfoundation.org/OFC/OFC-1-7.html

[lxviii] Fractal Foundation. Branching Fractals. Fractal Foundation. 2019.

https://fractalfoundation.org/OFC/OFC-1-1.html

[lxix] Fractal Foundation. Fractals in the body. Fractal Foundation. 2019. https://fractalfoundation.org/OFC/OFC-1-2.html

[lxx] Fractal Foundation. Fractal Blood Vessels. Fractal Foundation. 2019. https://fractalfoundation.org/OFC/OFC-1-3.html

[lxxi] Fractal Foundation. Fractal Neurons. Fractal Foundation. 2019. https://fractalfoundation.org/OFC/OFC-1-6.html

[lxxii] Picture 20 was retrieved from https://www.etsy.com/hk-en/listing/203934132/vintage-botanical-fern-temporary-tattoo

[lxxiii] Picture 21 was retrieved from https://packetpushers.net/make-like-tree-branch/

[lxxiv] Picture 22 was retrieved from https://fractalfoundation.org/OFC/OFC-1-2.html

[lxxv] Picture 23 was retrieved from https://pixabay.com/illustrations/fractal-spiral-endless-mathematics-199054/